The DelFly

G.C.H.E. de Croon · M. Perçin
B.D.W. Remes · R. Ruijsink
C. De Wagter

The DelFly

Design, Aerodynamics, and Artificial Intelligence of a Flapping Wing Robot

 Springer

G.C.H.E. de Croon
Delft University of Technology
Delft
The Netherlands

M. Perçin
Delft University of Technology
Delft
The Netherlands

B.D.W. Remes
Delft University of Technology
Delft
The Netherlands

R. Ruijsink
Delft University of Technology
Delft
The Netherlands

C. De Wagter
Delft University of Technology
Delft
The Netherlands

ISBN 978-94-024-1405-9   ISBN 978-94-017-9208-0   (eBook)
DOI 10.1007/978-94-017-9208-0

Springer Dordrecht Heidelberg New York London

Printed on acid-free paper

Springer Science+Business Media B.V. Dordrecht is part of Springer Science+Business Media
(www.springer.com)

Foreword

When I was young, I was fascinated by all flying things—ranging from small model helicopters to flying birds. At the time, the only model ornithopters were rubber band powered. After twisting the rubber band multiple times, such a lightweight balsawood ornithopter would fly for several metres, giving a brief impression of what an artificial flying insect would look like.

The extreme weight restrictions of such ornithopter models have considerably delayed the development of flapping wing vehicles with respect to rotorcraft or fixed wing vehicles. Only when the computer and mobile phone industry went sufficiently far in the miniaturisation of electronics and batteries, it became possible to create electrically driven flapping wing vehicles. The first flapping wings still quite resembled their rubber band ancestors, but along the line many new designs have sprung up. On the one hand, the on-board electronics allowed new actuator designs. For instance, the Vamp flapping wing toy I developed for a toy company was able to turn by changing the incidence of the wings. On the other hand, electronics allowed the further downscaling of the design. In 2007, I presented the 'Hummer' flapping wing at the IMAV in Toulouse, France. It weighed a mere 1 g.

Despite the technical advances and increasing scientific insights, flapping wing vehicles are still a rather uncharted domain. This book is the first of its kind to try and shed light on the ensemble of disciplines relevant to performing full missions with flapping wing vehicles. The DelFly flapping wing project is unique due to the many aspects that are studied, ranging from propulsion to aerodynamics and even artificial intelligence.

The reader will not only be impressed by the latest feats in autonomous flight and the latest insights into flapping wing aerodynamics, but will definitely profit from the experience and expertise conveyed by the DelFly team members. The book provides (sometimes surprising) insights in flapping wing design principles, but also gives very concrete pointers to the construction of a flapping wing vehicle with relatively cheap, commercially available components. Although the book is structured around the DelFly project, I would therefore recommend it to anyone who is interested in building or performing research on flapping wing vehicles on his or her own. In this way, it will definitely spur the further progress on flapping wing vehicles.

Petter Muren
CEO Proxdynamics

Preface

The DelFly project that is at the heart of this book has its origins in 2005 as a design synthesis project for students. The goal of the project was to 'design a flapping wing UAV of <50 g with on-board camera that will impress the jury of the European Micro Air Vehicle conference and competitions 2005'. After flying the 35 cm wing span and 21 g DelFly I at the EMAV 2005, winning the prize for 'most exotic MAV', the DelFly project was continued by the faculty of Aerospace Engineering of Delft University of Technology. It resulted in the creation of the Micro Air Vehicle Laboratory, which focuses on the design and study of all types of autonomous MAVs.

In 2006, with financial support of TNO (Netherlands Organisation for Applied Scientific Research), the DelFly II project was defined: this time the goal was to make a flapping wing MAV which would fit in a sphere with a diameter of 30 cm. The DelFly II was presented one year later in 2007. It surpassed the project's goals: besides reducing the wing span to 28 cm, the flight envelope was considerably increased. The DelFly II is able to fly forward at 7 m/s, hover, and even fly backward at -1 m/s.

DelFly II's broad flight envelope and on-board camera have made it a desirable study object both for investigating the airflow around the flexible wings and for achieving autonomous flight capabilities. The insight into the structural and aerodynamic properties of the DelFly II first led to the successful design of the DelFly Micro, presented in June 2008. The DelFly Micro is currently still the smallest (10 cm wing span) and lightest (3.07 g) flapping wing MAV in the world that carries both a camera and a video transmitter—a fact mentioned in the Guinness Book of Records 2010. Concerning autonomous flight, we have developed algorithms to continually increase the capabilities of the DelFly. We did not only apply these techniques to a laboratory setting, but have been demonstrating these techniques also in the IMAV competitions. The DelFly II was the first IMAV entry ever to perform autonomous flight indoors, successfully flying an 8-shape figure in the indoor dynamics mission at the EMAV 2008. At the IMAV 2010 in Braunschweig, Germany, the DelFly II was the only MAV that flew autonomously during the dynamics competition and it won the general first prize (beating all other types of MAVs) in the exploration competition. At the end of 2013 improvements to the motor and wings have allowed the design of the DelFly Explorer, the world's first fully autonomous flapping wing MAV. The DelFly Explorer carries a 4 g stereo vision system. It can take-off, keep its height, and avoid obstacles for as long as its battery lasts—with all sensing and processing performed on-board.

The design of an autonomously flying flapping wing MAV requires knowledge and expertise in various areas, including materials, aerodynamics, electronics, propulsion, flight control, and artificial intelligence. This book intends to convey the knowledge we gained in these areas to researchers, students, or enthusiasts that are interested in flying robots in general or flapping wing MAVs in particular. The main body of the book explains the scientific and engineering work performed to arrive at the current design and capabilities of the DelFly, always putting it in the perspective of other work on (flapping wing) MAVs. The in-depth chapters include a general introduction so that readers familiar with one of the mentioned domains will be able to follow the research in the other domains as well.

We hope that the ensemble of scientific, engineering, and practical insights contained in this book will further stimulate the research on flapping wing Micro Air Vehicles.

Delft   Guido de Croon
June 2015   Mustafa Perçin
                                                                                                         Bart Remes
                                                                                                      Rick Ruijsink
                                                                                          Christophe De Wagter

Acknowledgments

As mentioned in the preface, the DelFly Project started out as a student assignment, and over the years many people have made significant contributions. We are thankful for all of their work and constructive ideas. In particular, we would like to thank D. Lentink—one of the initiators of the project and a contributor to the DelFly I and II. Moreover, we are very grateful to B. van Oudheusden and H. Bijl from the aerodynamics department of TU Delft, with whom we have collaborated on all aerodynamics studies. Furthermore, we are grateful to the following persons from the aerodynamics department (B. Bruggeman, K.M.E. de Clercq, M.H. Groen), from the MAV-lab (N. Bradshaw, A. Koopmans, K. Lamers, T. Reichert, K. Scheper, S. Tijmons, J. Verboom), from the control group (S. Armanini, J. Caetano, C.C. de Visser, E. de Weerdt), and from the initial DelFly I student group (A. Ashok, K.M.E. de Clercq, D.A.J. van Ginneken, C.J.G. Heynze, S.R. Jongerius, A.N.A. Kacgor, R.C.A. Lagarde, P. Moelans, W.V.J. Roos, M.H. Straathof, G.J. van der Veen). In addition, we are grateful for the interesting discussions with P. Muren, CEO of Proxdynamics, and the feedback from V. Trianni, S. Tijmons and M. Karasek in reviewing chapters for this book. We would also like to thank M. van Tooren, B. Droste, H. Bijl, J.A. Mulder, M. Mulder and M.P. Oosten, who have been of a great support to our efforts.

Contents

Introduction

1

Abstract

This chapter sketches the dream of constructing small, intelligent flapping wing robots. The reader gets a first glimpse of the difficulties in understanding the aerodynamics of flapping wing robots. In addition, we touch upon the challenges of making these robots sufficiently intelligent to fly and navigate by themselves.

1.1 Flying like Animals

In the history of humankind, many people have looked at flying animals with reverence and envy. It is fascinating to see birds that hover in the air on a windy day, only to dive down when they spot prey. In the same vein, it is amazing to realize that tiny fruit flies successfully perform a rich repertoire of behaviors, ranging from the search of food to the social interactions involved in mating.

While a very old human wish is to fly like such animals, this book treats the slightly newer wish to construct a small machine that performs the same feats as flying animals. Two goals for such a small flying machine (or robot) are that it should have (1) similar flight characteristics to flying animals, and (2) the required intelligence for autonomously performing tasks in unknown environments. Below, we explain why these goals are not easy to attain.

1.2 Flight Characteristics

The flight characteristics of small flying animals strongly depend on the fact that they flap their wings in order to fly. The most enticing characteristic is perhaps the broad flight envelope of many flying animals: flapping allows them to perform both forward and hovering flight. A great example of this is the hummingbird, which can cover large distances during migration, but can also perform subtle maneuvers when feeding on a flower (cf. [15]).

The construction of a flapping wing robot is a challenging task, because there is still a relatively limited understanding of flapping wing propulsion in comparison with other propulsion mechanisms. The first attempts toward human flight involved flapping wings, but remained unfruitful. Instead, the first successful human flight was achieved by means of machines that did not exhibit flapping wing movements. The first sustained flights were performed in the 1800s with the hot-air balloons invented by Montgolfière. In the centuries that followed, progress was made towards other flying machines that also do not mimic the flapping wing flight of animals. The Wright brothers were the first to perform flights with a fixed wing aircraft in 1903 [10], while the first operational rotorcraft appeared around 1936. The increased insight into the aerodynamics of these flying machines actually resulted in more questions about how animals such as small birds and insects are able to fly; neither the surface of the wings, nor the speed of the flapping movements seemed to be sufficient under steady airflow conditions [1,3].

It was only in the 1980s that some initial insight was gained into flapping wing flight: animals exploit *unsteady* airflow in order to fly [4–8]. While unsteady airflow is problematic for fixed wing and rotary propulsion, it is the basic mechanism that— if modulated well—enables flapping wing propulsion. This insight has led to various studies of the unsteady airflow around flapping wings. The simulation of this airflow is still an active area of research, since the properties of the flow involve complex interactions between air and flexible structure in three dimensions (cf. [12]). As a consequence, simulators of unsteady airflow are still rather approximative and empirical research on flapping wings is performed to ameliorate them.

Studies on physical wings are either performed with flying animals or with artificial wings. Tests with flying animals have been very insightful, but are also limited. One limitation is that it is quite hard to train animals for the experiments. As an

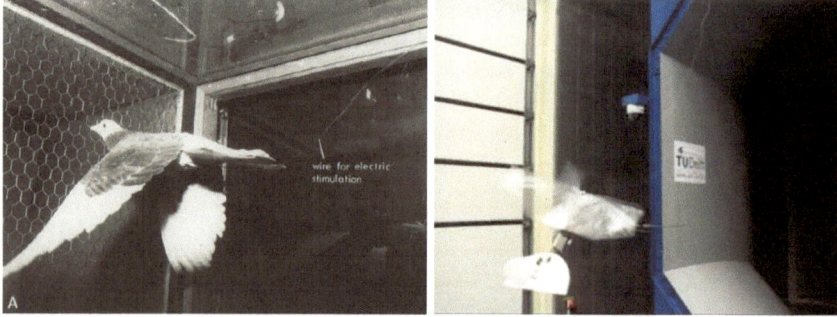

Fig. 1.1 *Left* A pigeon flying in a wind tunnel [14]. Reprinted with the kind permission of Springer Science and Business Media. *Right* A DelFly II flapping wing robot flying in the wind tunnel; with the help of external and onboard sensors it automatically flies at a specified place [18]. See https:// www.youtube.com/watch?v=Fi3ABv9E2Hw

example, in experiments with for instance pigeons [14], the animals have to be trained to fly in the test section of the wind tunnel (see the left part of Fig. 1.1). To stimulate this behavior, in [14] very mild electric shocks were used. The success and duration of training depended on the pigeon race. Some races are not so suitable for wind tunnel tests due to nervous flight behavior or their tendency to perform pronounced gliding phases. Another limitation of experiments with animals regards the relation between the wing structure, control, and resulting aerodynamics during flight. Surely all three components have undergone some form of optimization during evolution, but often it is unclear what the exact goals of the optimization were and what kind of restrictions have been imposed by nature. For example, in nature, a wing results from a growing process of different cells. The growing process most probably excludes a large set of possible wings that may have better characteristics concerning certain optimization goals.

Tests with artificial wings are not subject to the same confusion about optimization goals and restrictions. This allows the experimenter to explore the possible design space more thoroughly, for instance to study fundamental trade-offs in the design. Many tests are performed with artificial wings in isolation of a possible body (e.g., [11]). However, it is also relevant to study the wings in the context of a complete flying robot. The main reasons for this are that:

1. Insights into its aerodynamics can immediately lead to improvements of its design. This can lead to a higher payload capability or longer flying times.
2. Having a flying robot allows to perform measurements of its dynamics in flight, thus eliminating undesired influences from a tether or clamp on the force measurements.
3. Ultimately, the robot can be used for aerodynamic tests while in-flight. The robot can be programmed to perform the exact behavior requested by the researchers (see the right part of Fig. 1.1 for preliminary experiments in this area with the DelFly II [18]).

In short, the construction of a flapping wing robot can help us to better understand the aerodynamic mechanisms involved in flapping wing flight. Of particular interest are the fluid-structure interactions, whose importance is steadily growing with new data available in the literature. The gained insights can in turn lead to the construction of better flapping wing robots.

1.3 Intelligence

The intelligence of small flying robots is an evenly uncharted area of research and represents a tremendous challenge. The artificial intelligence for large driving robots is rather advanced. For example, the robotic cars in the 'grand DARPA challenge' 2005 autonomously drove a 212 km off-road track involving narrow tunnels and more than a 100 sharp left and right turns (see the left part of Fig. 1.2 for a picture of the

Fig. 1.2 *Left* 'Stanley', the autonomous robot that won the 2005 DARPA grand challenge [16]—public domain image from [2]. *Right* The DelFly Explorer, a 20 gram flapping wing robot that can fly fully autonomously [19]—public domain image from [17]

winning robot from Stanford, named 'Stanley' [16]). However, the successes of such robots rely on accurate and heavy sensors and on powerful computers. Going from ground robots in the order of kilograms and meters to small flying robots in the order of grams and centimeters, implies that the sensors become at best less accurate and at worst unavailable. Available computing power and energy also are dramatically reduced at such small scales. In the same time, small flying robots require faster responses from the robot, especially in confined spaces.

The hardware and dynamics of small flying robots require specific artificial intelligence techniques. Since nature is capable of solving similar tasks, it is not surprising that it has served as a main inspiration source for such techniques. In particular, computationally efficient optical flow algorithms and light-weight gyros have been employed to achieve autonomous flight in restricted environments [9,20]. However, in order to move beyond the current capabilities, biological inspiration may not suffice. For one, much of the natural flying intelligence is still not well understood. In addition, the different hardware used on flying robots may lead to the requirement of an artificial intelligence different from that of flying animals.

If one dreams of creating a tiny flapping wing robot that can fly autonomously, then having a physical and fully functioning robot is of great value [13]. Real-world tests force the experimenters to take into account all aspects of the robotic system. In addition, they reveal physical properties of the system that can sometimes be exploited. Finally, it also helps to identify the most important challenges for such robots.

To illustrate these points, please think of research on obstacle avoidance. There is ample research on obstacle avoidance assuming perfect knowledge on the obstacles' positions and velocities. However, implementation of the so-developed obstacle avoidance methods on a real flying robot will force the experimenter to face the uncertainties and limitations inherent to real actuators and sensors. The right part of Fig. 1.2 shows the DelFly Explorer, a 20-gram flapping wing MAV that can fly fully autonomously. It does so with the help of a 4-gram stereo vision system that uses an STM32F405 processor running at 168 MHz with a mere 192 kB of memory. Its obstacle avoidance strategy takes into account its limited field of view and its limitation by a maximal turn rate and minimal forward velocity.

1.4 Flapping Wing Micro Air Vehicles

We study the intricacies of small flapping wing robots by means of the DelFly, a Flapping Wing Micro Air Vehicle (FWMAV) created at Delft University of Technology. The DelFly project started in 2005 and has always focused on controlled-flight capable FWMAVs. All DelFly versions (see Appendix A) carry at least a camera on board. Over the years, we have studied various aspects of FWMAVs, ranging from design to aerodynamics and artificial intelligence. In this book, we will delve into these different topics.

Since we want the book to be suitable for students, engineers, and researchers from different disciplines, we will give general introductions to these topics, which are then followed by more detailed chapters containing research studies on the DelFly. The remainder of the book is organized according to the different aspects of the DelFly: its design (Chaps. 2–4), aerodynamics (Chaps. 5–6), and artificial intelligence (Chaps. 7–10). We conclude and reflect upon future work in Chap. 11.

References

1. R.J. Bomphrey, G.K. Taylor, A.L.R. Thomas, Smoke visualization of free-flying bumblebees indicates independent leading-edge vortices on each wing pair. Exp. Fluids **46**(5), 811–821 (2009)
2. DARPA. An official darpa photograph of stanley at the 2005 darpa grand challenge (2005), http://commons.wikimedia.org/wiki/File:Stanley2.JPG
3. R. Demoll, Zuschriften an die Herausgeber. Der Flug der Insekten und der Vgel. Die Naturwissenschaften **27**, 480–482 (1919)
4. C.P. Ellington, The aerodynamics of hovering insect flight. I. The quasi-steady analysis. Philos. Trans. R. Soc. Lond. B Biol. Sci. (1934–1990) **305**(1122), 1–15 (1984)
5. C.P. Ellington, The aerodynamics of hovering insect flight. II. Morphological parameters. Philos. Trans. R. Soc. Lond. B Biol. Sci. (1934–1990) **305**, 17–40 (1984)
6. C.P. Ellington, The aerodynamics of hovering insect flight. III. Kinematics. Philos. Trans. R. Soc. Lond. B Biol. Sci. **305**(1122), 41–78 (1984)
7. C.P. Ellington, The aerodynamics of hovering insect flight. IV. Aeorodynamic mechanisms. Philos. Trans. R. Soc. Lond. B Biol. Sci. (1934–1990) **305**(1122), 79–113 (1984)
8. C.P. Ellington, The aerodynamics of hovering insect flight. V. A vortex theory. Philos. Trans. R. Soc. Lond. B Biol. Sci. (1934–1990) **305**(1122), 115–144 (1984)
9. N. Franceschini, S. Viollet, F. Ruffier, J. Serres, Neuromimetic robots inspired by insect vision. Adv. Sci. Technol. **58**, 127–136 (2008)
10. R.G. Grant, *Flight: The Complete History* (DK Publishing, New York, 2007)
11. V. Malolan, M. Dineshkumar, V. Baskar, Design and development of flapping wing micro air vehicle, in *42nd AIAA Aerospace Sciences Meeting and Exhibit, 5–8 January, Reno, Nevada* (2004)
12. T. Nakata, H. Liu, A fluid-structure interaction model of insect flight with flexible wings. J. Comput. Phys. **231**(4), 1822–1847 (2012)
13. R. Pfeifer, C. Scheier, *Understanding Intelligence* (MIT Press, Cambridge, 1999)
14. H.-J. Rothe, W. Biesel, W. Nachtigall, Pigeon flight in a wind tunnel. J. Comp. Physiol. B. **157**(1), 99–109 (1987)

15. R.K. Suarez, Hummingbird flight: sustaining the highest mass-specific metabolic rates among vertebrates. Experientia **48**(6), 565–570 (1992)
16. S. Thrun, M. Montemerlo, H. Dahlkamp, D. Stavens, A. Aron, J. Diebel, P. Fong, J. Gale, M. Halpenny, G. Hoffmann, K. Lau, C. Oakley, M. Palatucci, V. Pratt, P. Stang, S. Strohband, C. Dupont, L.-E. Jendrossek, C. Koelen, C. Markey, C. Rummel, J. van Niekerk, E. Jensen, P. Alessandrini, G. Bradski, B. Davies, S. Ettinger, A. Kaehler, A. Nefian, P. Mahoney, Stanley: the robot that won the darpa grand challenge. J. Field Robot. **23**(9), 661–692 (2006)
17. C. De Wagter, Delfly explorer, a 20 gram flapping wing mav with stereo camera and onboard image processing to achieve autonomous collision free flight (2013), https://en.wikipedia.org/wiki/DelFly#/media/File:DelFly_Explorer_2013_V1.jpg
18. C. De Wagter, A. Koopmans, G.C.H.E. de Croon, B.D.W. Remes, R. Ruijsink, Autonomous wind tunnel free-flight of a flapping wing mav, in *EuroGNC 2013, Delft* (2013)
19. C. De Wagter, S. Tijmons, B.D.W. Remes, G.C.H.E. de Croon, Autonomous flight of a 20-gram flapping wing mav with a 4-gram onboard stereo vision system, in *2014 IEEE International Conference on Robotics and Automation (ICRA 2014)* (2014)
20. J.-C. Zufferey, *Bio-inspired flying robots: experimental synthesis of autonomous indoor flyers* (EPFL/CRC Press, Lausanne, 2008)

Part I
Design and Materials

Introduction to Flapping Wing Design

2

Abstract

This chapter treats the main choices, issues, and tradeoffs in the design of flapping wing MAVs. In particular, we discuss the implications of different tail and wing configurations, the energy source and various types of actuators. We also show how choices elementary to aircraft design, such as the trade-off between fuel/battery mass and payload mass can have rather large effects at the scale of light-weight flapping wing MAVs.

2.1 Introduction

The design of flapping wing MAVs is still a very active area of research. While the model plane world has known small rubber band powered ornithopters since the 1870s [4], the first electric powered flapping wing MAV, named the *MicroBat*, only flew in 1998 [28]. The design of flapping wing MAVs until now mainly progressed by means of trial-and-error. Automatic optimization is still very unreliable due to a lack of accurate theoretical models. Especially the design decisions concerning the shape, tension, and materials of the wings cannot be made purely on the basis of simulation due to a lack in knowledge on the aerodynamics around flexible airfoils.

Despite the lack of full theoretical grounding of all design choices, many functioning flapping wing designs have been made. Looking into these existing designs reveals some insights that may help to understand the key challenges and tradeoffs involved in flapping wing design. With these existing flapping wing MAV designs in mind, we discuss some of the main design choices and their consequences. We start with the general design concept in Sect. 2.2. Subsequently, the important choice of tail configuration is discussed in Sect. 2.3. This is followed by the wing configuration and single wing design (Sect. 2.4). We explain various methods to control the MAV and the possible implementations of such methods with actuators in Sect. 2.5. In Sect. 2.6 we discuss some of the choices that influence the energy and power

© Springer Science+Bussiness Media Dordrecht 2016
G.C.H.E. de Croon et al., *The DelFly*, DOI 10.1007/978-94-017-9208-0_2

available to flapping wing MAVs to perform their missions. Then, we touch upon the drive mechanism used to achieve flapping wing movements of the right frequency in Sect. 2.7. Finally, we draw conclusions in Sect. 2.8.

2.2 General Design Concept

An aircraft design is highly dependent on the intended use of the platform. The driving force in the design can be to optimize for maximum endurance on one hand, or on the contrary to optimize for minimum size. Typically the goal also includes other aspects such as stability or payload capability. Different combinations of goals can lead to very different designs. Common to almost all of these goals is that they are harder to attain at smaller scales. There are coarsely two approaches to finally arrive at fully functioning fly-sized flapping wing MAVs: bottom-up and top-down.

The bottom-up approach focuses on constructing and testing the tiny parts necessary for directly constructing a fly-sized robot [25,31,32,38]. Research studies adopting this approach often tackle extremely difficult sub-tasks such as the construction of the insect thorax [39], or the generation of sufficient thrust [2]. Some of the most ground-breaking early work was performed on the Micromechanical Flying Insect (MFI) [11]. In recent work, researchers from Harvard published on the first controlled flights of their fly-sized robot named *Robobee* [25,38]. The Robobee can be fully controlled, both attitude and position, by using control of the two fly-sized wings. In [25], the control relied on an external motion tracking system. However, later studies already used onboard sensors [12,30]. While the energy for flight is currently still provided externally via wires, the plans are to integrate the energy and also processing on board the flapping wing MAV.

The top-down approach starts with relatively larger scale but fully functioning flapping wing MAVs (e.g., [7]. The idea behind the approach is that studying such MAVs can lead to insights for the construction of a following, smaller or smarter version. One advantage of this approach is that it allows interplay between theory and practice. For aerodynamics research, having a flying system ensures that the research is directed to aspects that also have a practical relevance. For artificial intelligence research, having a physical and fully functioning MAV is of great value: real-world tests force the experimenters to take into account all aspects of the robotic system. In addition, they reveal physical properties of the system that can be exploited by the algorithms.

The bottom-up and top-down approach have complementary advantages and risks. For example, a risk of the top-down approach is that the research will focus too much on incremental modifications of the MAV, ignoring possible disruptive improvements. On the other hand, a risk of the bottom-up approach is that research may focus too much on detailed aspects that might turn out irrelevant for a fully flying system. The bottom-up approach can lead to fundamental new understanding and techniques, while the practical 'surprises' of the top-down approach give insight into pressing problems of lacking scientific knowledge or technology. We believe that progress in flapping wing research requires both approaches to exist.

In the DelFly project a top-down approach was adopted, because of our interest, expertise, and means. In what follows, we will mostly limit ourselves to fully functioning MAV designs, carrying their energy source on board.

2.3 Tail Configuration

Perhaps the most influential design choice of a flapping wing MAV is its tail configuration. A tail damps the rotational dynamics, implying that around the nominal flight condition the flapping wing MAV has a passively stable attitude. Since a tailed design does not necessarily need active control for stabilizing the attitude, there is no need for an onboard Inertial Measurement Unit (IMU) or autopilot. In addition, directional control can be achieved with parts of the tail, as is done with normal fixed wing aircraft (see Sect. 2.5). The wings do not have to be used for directional control and can be realized with relatively straightforward mechanisms. Figure 2.1 shows different possible tails, including a conventional plane tail (I - from [17]), inverted V-tail (II - from [7]), and a tail that also serves as a landing gear (III - from [7]).

A tailless design is closer to the anatomy of flying insects, but makes the platform's attitude passively unstable [19,33,34]. As a consequence, a high-bandwidth control system needs to act continuously in order to stabilize the attitude. How to achieve

Fig. 2.1 Different tail designs: (*I*) conventional plane tail, Wright State University flapping wing MAV [17], (*II*) inverted V-tail, DelFly I [7], and (*III*) 'standing tail', DelFly II [7]. Images reprinted with permission

Fig. 2.2 Nano
Hummingbird tailless design
[20]—public domain image
from [6]

this with the various degrees of freedom of flapping wings was for a long time an
open question. A control system can for example change the flapping frequency of
the individual wings, the phasing of their flapping cycles, the flapping amplitudes,
possibly the angles of the wings during the flapping cycle, etc. There was already
theoretical work on the principles that could establish attitude stabilization (e.g.,
[9,10]). However, the first design actually realizing controlled flight of a tailless
MAV was the Nano Hummingbird [20], shown in Fig. 2.2.

2.4 Wing Configuration and Design

Based on the tail configuration, a specific wing configuration can be chosen.
Figure 2.3 shows a number of such configurations. The most 'traditional' designs
are perhaps the ones with single wings (e.g., as in the 'Small bird' / 'Big bird' [15]).
Figure 2.3 I shows the design of the 'Robo Raven', which can actuate its left and
right wings independently of each other. Another bio-mimicking design is that with
two wing pairs behind each other, as in the dragonfly (Fig. 2.3 - II) [13]. Going
beyond nature, there are also designs that feature four wings with the same stroke
plane (e.g., [1,7,17,24,41] (Fig. 2.3 - IIIa/b/c). The wings of these designs typically
almost touch each other at one or more points during the flapping cycle. When they do,
they first 'clap' together and then 'fling' apart, providing additional lift (see Chap. 5).
The wings can perform a single clap-and-fling (IIIa), a double clap-and-fling (IIIb),
or multiple clap-and-flings simultaneously (IIIc) both during the outstroke and the
instroke. Obviously, the wing configuration has large consequences on the forces
generated by the flapping and on the way in which the MAV can be controlled.

After choosing a wing configuration, there still remain endless probabilities for
designing the individual wings. The wing design is very important for the aerodynamic performance, and hence the lift that can be generated by the flapping wing

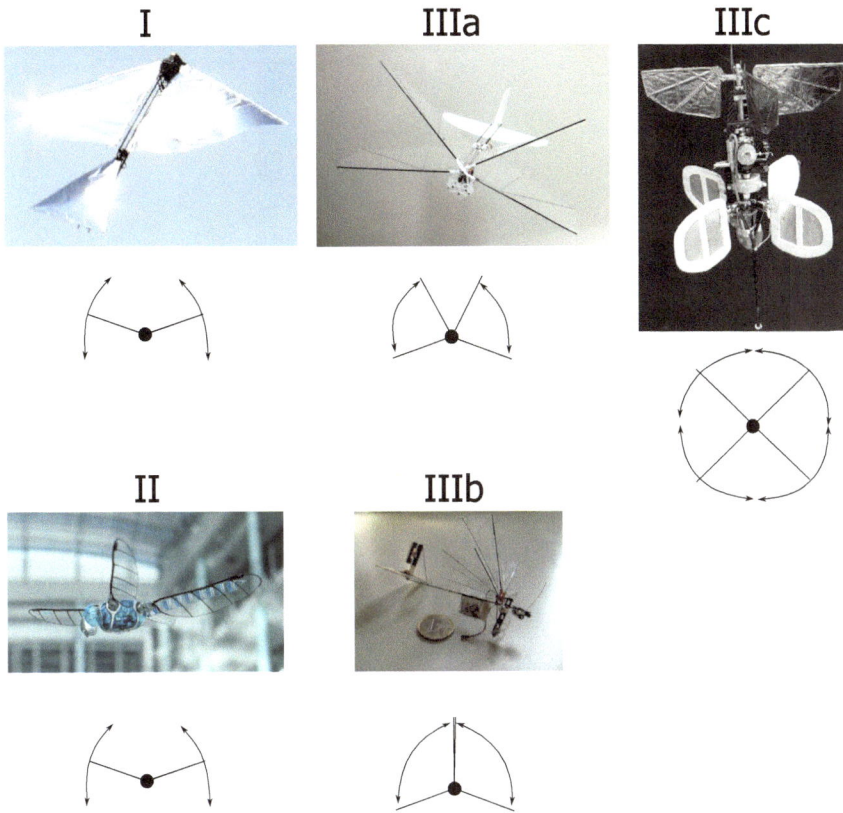

Fig. 2.3 Different wing configurations: (*I*) the 'Robo Raven' [14], (*II*) dragonfly setup of the 'Bionicopter' [13], (*IIIa*) DelFly II, (*IIIb*) DelFly Micro with double clap-and-fling (public domain image from [37]), (*IIIc*) the 'Mentor' [41]. All images reprinted with permission

MAV. It involves a choice of the materials for the structuring elements and wing membrane, and the way to combine these to form the wing's shape and structure.

A traditional choice for the wing materials consists of PET-foil and carbon fiber reinforced polymer (CFRP) rods. These materials have proven their worth, are widely available, and do not require specific infrastructure for construction. The downside is that they typically require some manual work, which can limit repeatability. Moreover, the design options with these materials are relatively limited. Most designs with PET-foil and rods are limited to geometric shapes with a stiff leading edge and a few stiffeners added to the wing. In order to allow for more intricate, and yet repeatable designs, other materials and fabrication methods have been investigated. For example, the early MicroBat project involved the creation of MEMS wings in various shapes [28]. In a more recent study, a complete flapping wing MAV has been 3D-printed [29]. Although these methods are very promising, they still face challenges concerning fatigue and lifetime.

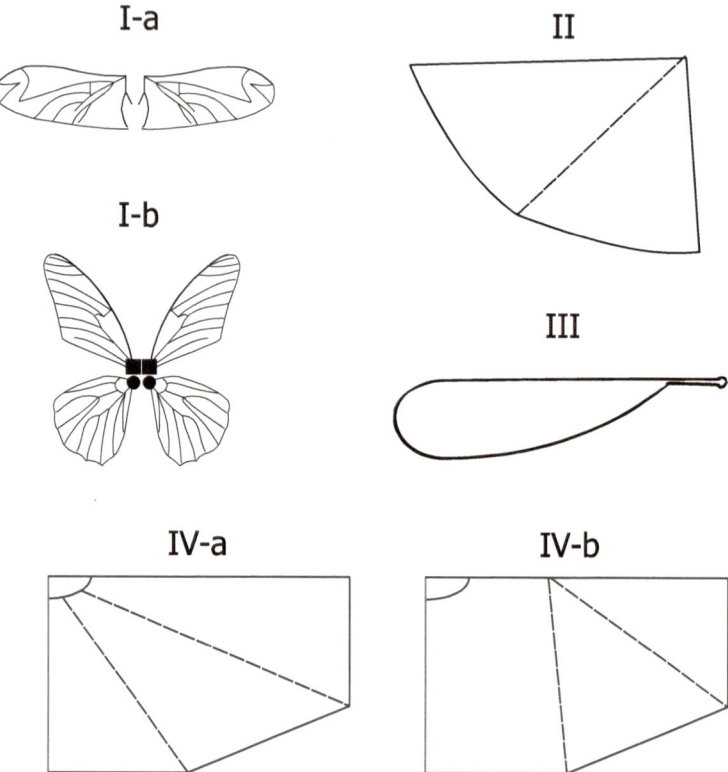

Fig. 2.4 Different wing designs: (*I-a*)/(*I-b*) sketches of the designs tested out for the MicroBat design [28], (*II*) sketch of the wing design of 'Small Bird' [3], (*III*) sketch of the 3D-printed wing by [29], and (*IV-a*)/(*IV-b*) sketches of DelFly II wings [8]

Based on the selected materials and fabrication method, a wing shape can be envisaged. Figure 2.4 shows different wing shapes from a number of studies [3, 8, 28, 29]. Wing designs I-a and I-b are examples of bio-mimicking wing designs, here copying the structure of a beetle and butterfly wing, respectively. The 'biomimicking' wings are often not the optimal ones for flapping wing MAVs (as was also the case in [28]. Wing design II, of the FWMAV 'Small Bird' [3], is perhaps closest to the rubber-powered ornithopters mentioned in the introduction. It uses a 'round' trailing edge. Wing design III is used on the 3D-printed flapping wing MAV presented in [29]. It has a 'round', solid outlining. It lacks stiffeners on the inside of the wing. Wing designs IV-a and IV-b use traditional PET foil and carbon stiffeners, but they do not have the round edges in order to improve the wing's efficiency [7]. The difference between wings IV-a and IV-b lies in the positioning of the stiffeners, leading to a significant difference in aerodynamic performance [8].

2.5   Control and Actuators

2.5.1   Actuation Strategies

Natural fliers use several strategies to control the flight. Whereas birds use their tails, flies rely mostly on their wings (and in some cases their legs). Also flapping wing MAVs can use the tail and / or the wings for control.

For tailed FWMAVs, several different actuation schemes have been devised, of which some examples are shown in the top row of Fig. 2.5. In the figure, the actuated elements are colored in orange. A common design is to have a conventional aircraft tail (design I in Fig. 2.5) with an elevator and a rudder. The elevator induces a pitching moment (see Fig. 10.2 for the definition of rotations and axes). The rudder has coupled effects. It initially induces a yaw moment, which in turn causes rolling. Design II shows a design as used on Wowwee's Flytech Dragonfly. It yaws by means of a tail rotor, just like a helicopter. Another option is to use a ruddervator, shown as design III in Fig. 2.5. This setup involves two actuated control surfaces on an inverted v-tail. The inverted v-tail has several advantages. For instance, it uses fewer tail surfaces, leading to lower interference drag and construction and weight advantages. Moreover, the tail produces a combined yawing and rolling moment that support each other. A drawback

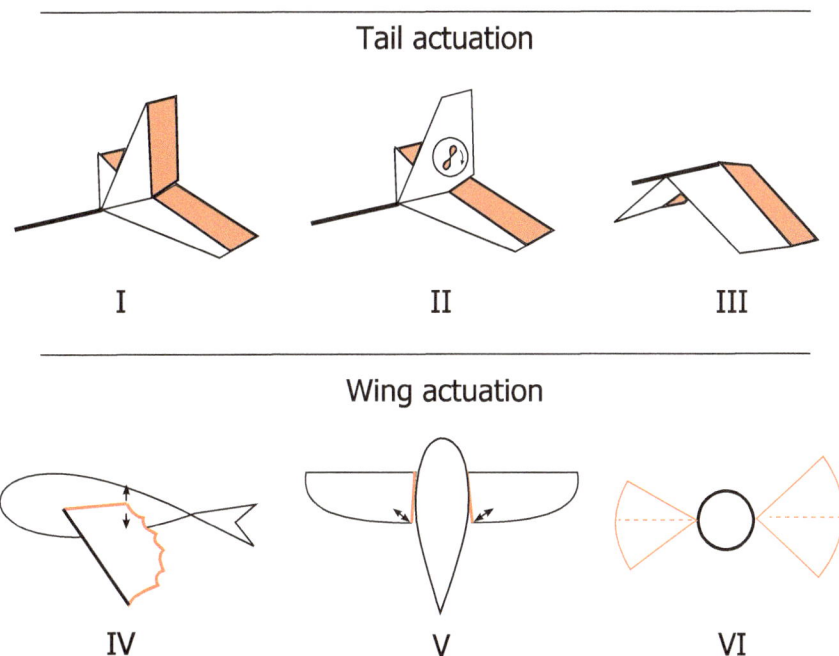

Fig. 2.5 Various actuation methods: (*I*) aircraft tail, (*II*) aircraft tail with a propeller for yawing, (*III*) inverted V-tail, (*IV*) changing the incidence of the wings in forward flight, (*V*) tensioning the wings for tailless hover flight, and (*VI*) changing the stroke amplitude and mean stroke position

arises from the coupled effects of actuating the control surfaces. For instance, a yawing deflection reduces the maximum attainable pitching deflection. In addition, the inverted v-tail makes landings more difficult - more easily leading to damage.

In contrast to birds, insects use their wings for both propulsion and control, which provides them with significant control authority and allows for aggressive maneuvers. The actuation design IV shown in Fig. 2.5 has been designed for the Vamp / Wasp toys. Instead of using the tail, the incidence of each wing is changed to introduce a yawing action that will result in turning. In order to turn, one wing is given a higher incidence than the other. The higher incidence wing will be subject to more drag, which causes the ornithopter to yaw. In the same time, the higher incidence wing will generate more lift, which will make the ornithopter roll slightly in the opposite direction. This control for turning works reasonably well at low speeds at which the aircraft is trimmed: the ornithopter will turn in the direction of the higher incidence wing. At higher speeds the adverse yaw effect diminishes and the roll effect starts to dominate. As a consequence, the control action will reverse, with the ornithopter turning toward the lower incidence wing. Hence, this control scheme is a problem if the ornithopter has to fly at very different speeds.

The actuation scheme V in Fig. 2.5 is used on the Nano Hummingbird [20]. The bars allow to tension / relax the wings, which will increase / decrease the lift generated by the wing. If the left wing is tensioned more than the right wing, a roll moment will be created. If both wings are tensioned more during the aft part of the flap stroke (behind the MAV), then a forward pitch moment will be created. A yaw moment is created when the wings are tightened asymmetrically, for one wing during the forward motion and for the other wing during the backward motion of the flap stroke. Alternatives to this method for hovering single wing MAVs have also been investigated. In [18] the flapping amplitude and the mean position of the flapping stroke are changed (method VI in Fig. 2.5). Increasing the flapping amplitude results in an increase in lift, while changing the mean position of the flapping stroke can create a pitch or yaw moment. A similar strategy is employed in the Robobee [25].

2.5.2 Actuators

In order to implement a control scheme, an actuator is necessary. Various actuators that are available (shown in Fig. 2.6) will be discussed here with their relevant properties, including mass, size, force, and speed.

Magnetic actuators (Fig. 2.6, I) are pulse width modulation (PWM) driven with a duty cycle proportional to the transmitter control stick, which results in a proportional current. This current produces a moment in the magnet which again translates to a proportional force that goes to the control surface. However the force is small and with the air pressure on the control surfaces being proportional to the air velocity squared, the control throw gets much lower at higher airspeed. This can cause inability to pull up from fast descending flight.

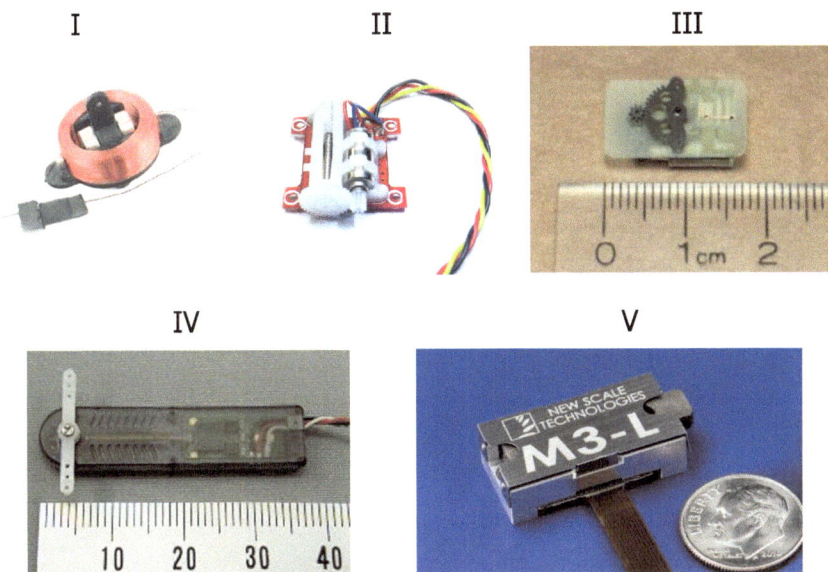

Fig. 2.6 Various actuators: (*I*) magnetic actuator by Plantraco [27], (*II*) conventional servo by Hobby King [21], (*III*) conventional servo by Microflierradio [23], (*IV*) Muscle wire by Toki ([5]), and (*V*) Piezo servo by New Scale [35]. Images reprinted with permission

An alternative consists of conventional servos (Fig. 2.6, II and III) that use a small electric motor, gearing and a potentiometer or magnetic-Hall sensor for position feedback. This type of actuator has more force and a higher accuracy compared to the magnetic actuator. Two example servos are shown in Fig. 2.6. Number II is a linear servo weighting only 1.1 g; it is industrially produced and marketed by Hobby King under the name Ultra Micro. The Ultra Micro is very close to a conventional servo but designed with size and weight reduction in mind. The linear potentiometer for feedback is also used as an attachment for the mechanical output. Even lighter designs are available, such as number III in Fig. 2.6, which is a rotary servo produced by Micro Flier Radio weighing 0.45 g.

Muscle wire is another option (Fig. 2.6, IV). It contracts when it is heated and relaxes when it cools down again. This is used to actuate a pivot where two wires work in conjunction. Because the contracted / heated wire has a lower resistance, a position feedback based on the differential resistance is employed for the servo action. The initial movement from cooled condition is relatively quick, while the relaxation to the neutral position is much slower. The cooling of the wire is limiting the actuation speed. For instance, when used with the DelFly, the speed was insufficient and the wires a bit too fragile.

Finally, Piezo-based servos can be used (Fig. 2.6, V). For instance, the New Scale M3-L is currently the smallest Piezo-based servo that could be used in small UAVs. Intrinsically the piezo servo can be very light, the heart of it measures only $2.8 \times 2.8 \times 6$ mm, but the high voltage electronics to drive the actuator and incorporate feedback

Table 2.1 Different actuators and their properties

		Magnetic actuator	Ultra micro servo	Bio wire servo	Linear servo
Manufact.		Plantraco	Microflier	Toki	HK UM
Mass	gr	0.7	0.45	1.0	1.1
Size	mm	10 × 7	12 × 10 × 6	38 × 9 × 3	18 × 15 × 8
Force	N	0.01	0.15	0.15	0.35
Stroke	mm	± 4	± 4	± 4	± 4
Speed	sec	0.1	0.15	+0.3 −0.9	0.18
Power idle	mA	0	5	5	5
Power Avg	mA	20	30	30	30
Power max	mA	80	100	80	120
		Micro servo HS5035HD	Piezo servo M3-L	Bare piezo plus electronics	
Manufact.		HiTec	New scale	New scale	
Mass	gr	3.6	4.5	0.8	
Size	mm	18 × 16 × 8	27 × 13 × 8	18 × 13 × 6	
Force	N	7.00	0.20	0.20	
Stroke	mm	± 6	± 3	± 3	
Speed	sec	0.12	0.6	0.6	
Power idle	mA	5	50	50	
Power Avg	mA	30	70	70	
Power max	mA	250	130	130	

with robust enough output still increase the size and mass to such values that the use in MAVs is not yet practical. Moreover, the servo is currently still relatively slow and the piezo drive requires high voltages. The implementation of piezo actuators would require and justify a specific research project on its own.

Table 2.1 summarizes the major specifications of seven different actuators.

2.6 Energy and Power

Energy and power are crucial parts of the flapping wing MAV design. The available energy heavily influences the flight time. Power, which is the amount of energy used per second, determines whether flight is possible in the first place. In addition, it is an important factor for determining how maneuverable the MAV is. Both energy and power impact the type of payload that can be carried aboard the MAV. The three characteristics of flight time, maneuverability, and payload capability are essential to the utility of an MAV's design and often have to be traded off against each other. For instance, if the payload uses more power, then a given energy source might just have enough power to fly but the climb rate might have become ridiculously small. And if more power is used, due to battery characteristics a same battery typically will provide less energy due to higher internal losses. In this section, we will explain some of the main factors influencing the mentioned three characteristics. In Subsection 2.6.1 we discuss flight efficiency and its dependence on the flight regime. Subsequently, in Subsection 2.6.2 we explore different energy storage materials and evaluate their promise and current applicability to light-weight flapping wing MAVs. Afterward, we focus on batteries and explore the trade-off between battery mass and payload mass in Subsection 2.6.3.

2.6.1   Flight Efficiency

Micro aircraft are not as efficient as their larger counterparts. At a small size or low velocity the viscosity of the air has a greater influence on the air flow. This low Reynolds number condition generally leads to higher drag and lower lift, implying less efficiency. The efficiency of flight is also related to the way in which the micro aircraft flies. Flapping wing flight exploits the viscous aerodynamic effects and hence partly overcomes the loss of lift. Furthermore, the flight mode of the flapping wing aircraft is of importance. For instance, hovering flight is a quite power intensive flight mode. This is also the case for animals: when hummingbirds hover, they need to feed very often (e.g., every 4–5 min) to stay airborne [16]. Likewise, in hover mode a flapping wing MAV will not be at its most efficient regime and will need more power and will show shorter flight times than at more efficient forward flight speeds.

2.6.2   Energy Storage Materials

When considering the qualities of different energy sources we look at the specific energy and specific power of the pure energy storage material. The specific energy measure of highest relevance is the energy-to-weight ratio (kJ/kg), although the energy density (kJ/liter) is also of importance. This also goes for the power-to-weight ratio (kW/kg) and power density (kW/liter). For a choice of energy storage material, we also need to take into account that the energy content of the storage material has to be converted to useable power for the aircraft. The aircraft needs mechanical power for propulsion and electric power for the flight control and payload systems. To convert the potential energy from the source to propulsive and electric power, a conversion system is needed. This conversion system can be simple and efficient for some energy sources while it can be completely impractical for others. The illustrative overview in Table 2.2 is derived from generally available information and can be regarded as representative for the energy carrier but not necessarily absolutely accurate for each application.

Table 2.2 Different energy sources and their properties

Storage material	Energy type	Energy-to-weight ratio kJ/kg	Power-to-weight ratio kW/kg
Uranium (in breeder)	Nuclear fission	80600000000	
Hydrogen (compressed at 70 MPa)	Chemical	142000	
LPG (including Propane /Butane)	Chemical	46400	
Gasoline (petrol) / Diesel / Fuel oil	Chemical	46000	
Jet fuel	Chemical	43000	
Fat (animal/vegetable)	Chemical	37000	
Coal	Chemical	24000	
Carbohydrates (including sugars)	Chemical	17000	
Protein	Chemical	16800	
Wood	Chemical	16200	
Formic Acid	Chemical	6100	
TNT	Chemical	4600	
Gunpowder	Chemical	3000	
Hydrogenperoxide	Chemical	2600	
Lithium SOCl2 (primary)	Electrochemical	1800	0.10
Hydrogen Fuel Cell, Medium size, Horizon Aeropack	Electrochemical	2400	0.12
Formic Acid + Fuel Cell System, Small, Neah Power	Electrochemical	1500	0.10
Lithium SOCl2 (hi-current, primary)	Electrochemical	1140	0.21
Lithium-Sulphur (secondary) (SotA: 2015)	Electrochemical	900	1.13
Lithium-ion -polymer battery	Electrochemical	650	2.80
Small Lithium-polymer battery (< 0.3 Ah)	Electrochemical	450	2.80
Alkaline battery (primary)	Electrochemical	670	0.06
Nickel-metal hydride battery	Electrochemical	360	0.60
Lead-acid battery	Electrochemical	170	0.10
Supercapacitor	Electrostatic	18	0.50
Electrostatic capacitor	Electrostatic	0.36	30.0

The energy sources in the "chemical" category have very interesting energy densities. For instance, hydrocarbon fuel engines, with internal combustion or jet turbines, are widely used in larger unmanned aircraft. These engines often drive a propeller and an electric generator for electric power to the payload, control and navigation systems. For micro or nano aircraft existing versions of these systems are too large to be accommodated.

Fuel cells that convert the chemical energy to electric power are still novel technology even for the larger unmanned aircraft. These systems are rather complex and only deliver a moderate amount of power, but they can do this for a long time. This restricts the use to efficient long endurance missions in the order of hours. A hybrid power source with a small battery that is constantly topped off can improve the versatility of the fuel cell system. For example, it can be used to deliver a short burst of power during take-off and short climbs. Small commercial systems in this category have a weight of about a kilogram and consequently the minimal mass of the complete UAV with fuel cells will generally not be less than about 2 kg. These systems attain already a very good energy-to-weight ratio, with a power density that is compatible with longer duration flights.

http://Neahpower.com is working on a system that uses formic acid as a source of hydrogen that is easier and safer on small scales than pressurized hydrogen. This allows the fabrication of smaller fuel cell systems for smaller UAVs. For MAVs the fuel cell technology is not yet suitable. Although the energy-to-weight ratio of a fuel cell system can be high, their power-to-weight ratio is low, and the high minimum system weight makes them inadequate for the very small MAVs.

There are UAVs flying on prototype Lithium-Sulphur cells made by the company SionPower. The cells have the potential of outperforming the Lithium Polymer battery by a factor of two, but the research progress has been much slower than anticipated. The cells are not yet available commercially but are tested in military and research vehicles like the Qinetiq Zephir. At this stage of development the Li-S cell is 50 % better than the Lithium Polymer in energy-to-weight ratio and a little worse in power-to-weight ratio. In the future this type of battery could be useful when the pace of development proceeds.

The possible use of solid (rocket) fuel has been developed to power small insect like gliders [22]. It could be very interesting for specific types of aircraft, but impractical for the majority of UAV uses due to short burn times.

Promising research takes place on ultra-small catalytic based internal combustion type of actuation for the flapping of the wings. Hydrogenperoxide is injected into a cylinder and is decomposed through the use of a catalyst [36]. This propulsion might be useable in the future for very small, micro or nano flapping wing air vehicles. Electric power for control, navigation and intelligence is still required.

For small to very small aircraft the electric power source in the form of a battery powering an electric motor yields a very interesting compromise between duration and complexity. The energy-to-weight ratio, the power-to-weight ratio and the power density of modern Lithium Polymer batteries is good. Even very small brushless electric motors reach a high power-to-weight ratio of over 1 kW/kg together with a good efficiency and controllability. Therefore, in the following we will limit our discussion to the use of batteries and electric motors.

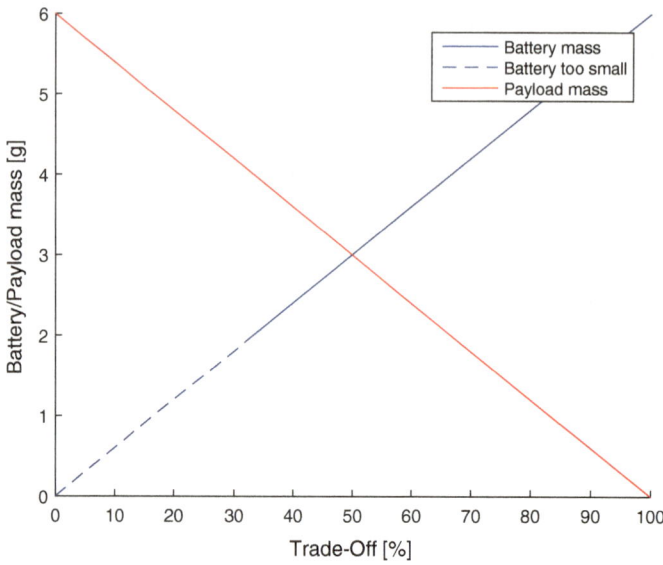

Fig. 2.7 Battery mass versus payload mass

2.6.3 Trading Off Battery Mass and Payload Mass

Given an energy source, a choice has to be made concerning the amount of energy to incorporate into the design. This choice is strongly related to the MAV's envisaged mission and is based on the trade-off between payload, maneuvering capabilities and flight duration. A given aircraft can lift a limited payload mass plus energy source mass with a given amount of maneuverability. Payload mass can be traded off with battery mass as shown in Fig. 2.7. When all the mass is used for energy the duration is maximized at the expense of payload. The minimum mass of the battery is a more complex value and is strongly influenced by the available power. When the energy source becomes small, although it might have sufficient energy for a very short flight, it can not always deliver its energy content sufficiently fast. In other words, the power-to-weight ratio of the battery will typically determine how small a battery can become to still support flight.

It is good to realize that although Fig. 2.7 shows a nice linear trade off in terms of weight, the consequences of choosing a battery and payload mass are not as linear at all. At smaller battery sizes the load factor will significantly increase, which can reduce flight time more than linearly.

Importantly, each battery has its associated (nonlinear) discharge curves (shown for the cyclonE-130 in Fig. 2.8). The discharge curve is determined by the amount of current (A) drawn from the battery. The amount of energy actually used for powering the MAV is represented by the area under each discharge curve (V × mAh). Relative load is expressed in 'C' and for a 130 mAh cell a discharge rate of 1 'C' means

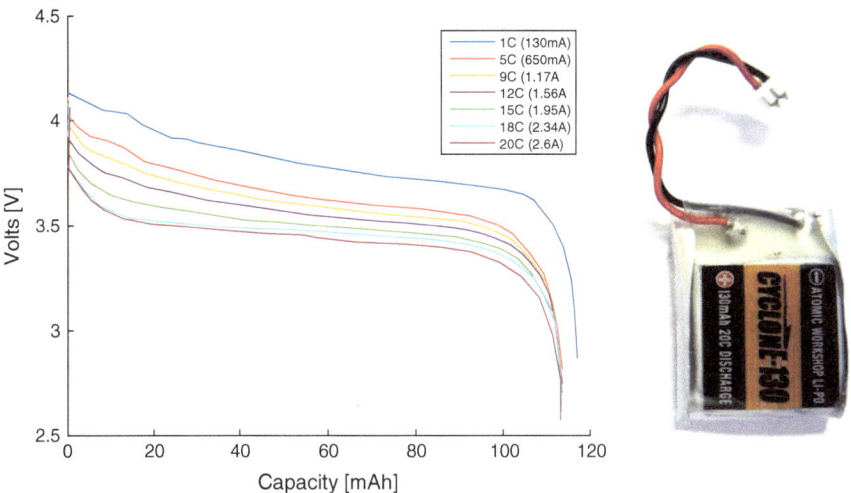

Fig. 2.8 *Left* Discharge curves of CyclonE-130 mAh cell. *Right* CyclonE-130 mA cell with Molex connector [40]

discharging at 130 mA. As can be seen from the plots, the area under the curve gets significantly smaller at higher relative loads.

Figure 2.9 shows the evolution over time of battery voltage at various discharge rates, and the bottom plot illustrates what happens with the lost energy. At higher discharge rates, the temperature rises significantly and further increasing the discharge rate even poses overheating risks.

The efficiency of a battery and also its maximal load depend mainly on the internal resistance of the cells used. Internal resistance is not a clearly-defined and easy-to-measure value but rather a complex chemical process in function of time. It is typically well approximated by measuring the voltage change due to a predefined reproducible load change using Ohm's law.[1] For instance during discharge, the load is dropped completely during 2 seconds and the voltage rise is measured after 2 s of cell relaxation. Figure 2.10 shows the discharge curves of a 150 mAh cell and the bottom plot shows how the internal resistance varies during the discharge. A few important things can be noted from this plot. First of all the internal resistance significantly increases when the cell is getting empty. At first sight it might look like higher loads on the battery reduce its internal resistance. In fact it is the higher temperature which is responsible for slightly lower internal resistances. The colder the lithium battery, the higher the internal resistance.

Figure 2.11 computes how much energy the same 150 mAh lithium-polymer battery can provide until its voltage drops below a certain voltage threshold. DelFly for instance has a small Electronic Speed Controller (ESC, used for brushless motors)

---

[1]$I = V/R$, with I current, V voltage, and R resistance.

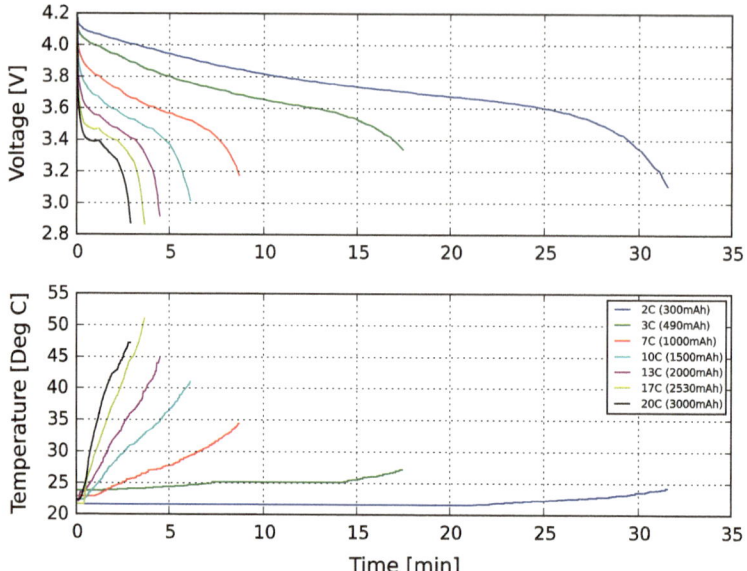

Fig. 2.9 Discharge curves of 150 mAh cell

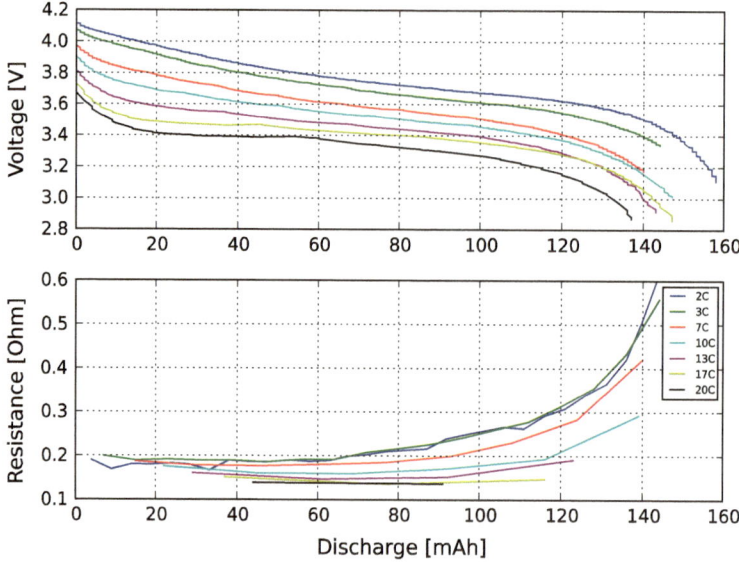

Fig. 2.10 Cell characteristics 150 mAh cell

that will function properly down to 3.3 V but not lower. This means that we should only consider the area under the discharge curve for the part in which the voltage is above 3.3 V. The tested cell from Fig. 2.11 would deliver only 0.33Wh until 3.3 V

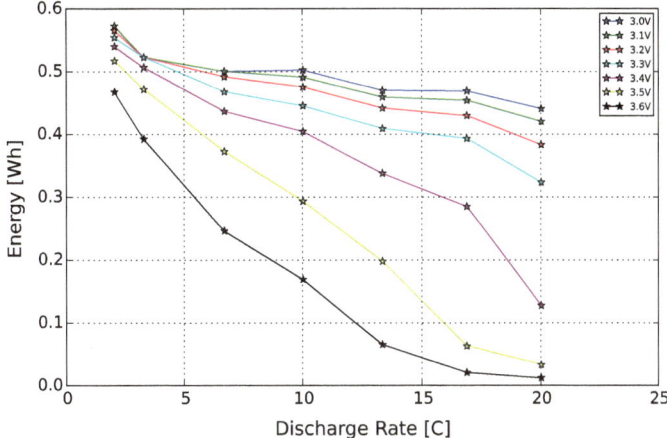

Fig. 2.11 Useful energy content 150 mAh cell

at 20C compared to 0.55 Wh at 2C, meaning a 40 % reduction in energy content. In other words, if the battery load is 10 times higher, the flight is not 10 times shorter but almost 17 times shorter. Similarly, if for instance an initial battery mass of 6 g is traded for 2 g of additional payload (See Fig. 2.7) resulting in only a 4 g battery, then one might expect $\frac{1}{3}$ less flight time. However, since the battery is 33 % smaller the relative load of the smaller cell is 50 % higher and so are internal losses. If the low voltage threshold is fixed, the higher load over the battery internal resistance also means the low voltage threshold is reached before the battery is actually empty. Figure 2.11 illustrates that this process can get quite dramatic.

The other way around, when for instance wing efficiency is increased and a lower power is needed, then the lower relative load on the cell will also mean a reduction of losses in the battery, and more useable energy before the low voltage threshold is reached. This is why a 10 % wing efficiency increase can yield more than 10 % extra flying time. This non-linearity is particularly large for very highly loaded or in other words small batteries.

In summary, small changes in mass or efficiency of the propulsion system can have a large impact on the flight duration due to the nonlinear discharge curves and requirements of the electronic components.

2.7   Drive and Mechanism

In most small ornithopters the wings oscillate with a frequency of between 8 and 50 Hz, depending on vehicle size and flight regime. To produce this motion one could employ actuators that directly produce a reciprocating movement. A linear or circular electromagnetic actuator like that of the magnetic pick-up head in a hard disk could be used. The problem is that this will be rather heavy. The actuator has to generate a significant force at a relatively low frequency. As the mass of any electromagnetic system is roughly proportional to the force it can produce, we can see that a high force at a low frequency will result in a heavy actuator.

Electrostatic actuators like those of the piezo type suffer also from a frequency mismatch. These systems perform best at really high frequencies in the order of 50–150 kHz and at low amplitudes. In systems that operate at 8 to 50 Hz the power-to-weight ratio of a piezo actuator is low. The Robobee flies on the basis of a Piezo actuator, flapping at 120 Hz [25]. However, the system currently still needs an external power source, with the power-to-weight ratio inadequate for untethered flight.

A rotational electric motor, also at the scale of micro and nano air vehicles, delivers a good power-to-weight-ratio of around 1 kW/kg at efficiencies of at least 60 % for micro to 40 % for the nano systems. Sensorless, brushless motors are possible at these scales and offer a superior reliability, power-to-weight ratio and efficiency compared to the brushed motors. The control of the sensorless brushless motor is not trivial but the required computing power is low compared to the total system power consumption.

Brushless motors consist of a permanent magnetic component and of a 'stator' with electromagnetic coils. The electromagnetic field can be controlled to attract the permanent magnetic components at the right time, making the permanent magnetic component spin around. Two types of brushless motors are known, so called inrunners and outrunners (inset I and II, respectively in Fig. 2.12). With the first type the stator containing the electromagnetic slots is positioned on the outside of the permanent magnetic armature that spins inside of the motor. The outrunner type has a central stator with the electromagnetic slots where the permanent magnetic rotor spins around the stator. The outrunner is preferred in our type of use as it normally produces more torque at a lower rpm (rotations per minute) than the inrunner. The optimal speed range of small outrunners is around 20.000 rpm. This circular

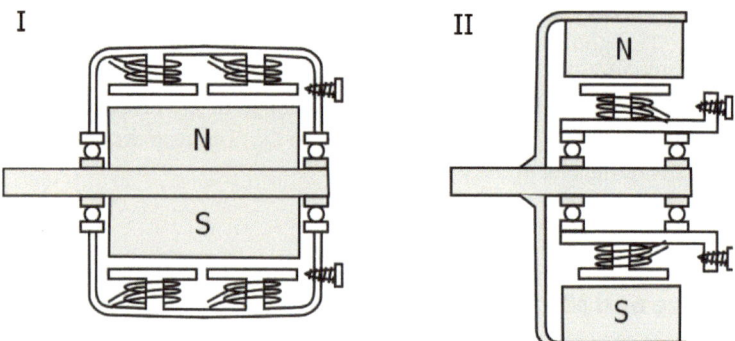

Fig. 2.12 I: Inrunner brushless motor. II: Outrunner brushless motor. In both cases the magnets are moving (*grey*) and the coils are static. The inrunner has the magnets at the centre while the outrunner has the magnets around the coils. The inrunner has a static outer case, while the outrunner must be mounted at the stator end, and care must be taken that nothing can touch the rotating outer hull

movement has to be transformed into a reciprocal movement of the right frequency. For instance, the DelFly has a two-stage gearing with a gear ratio of 21.3:1. Therefore, at a motor speed of 19.200 rpm we achieve a representative 15 Hz flapping frequency.

Extreme examples of micro flappers include the extraordinary 1 g 'Hummer' [26] (a flapping wing model that served as the basis for the 3 g 'Da Vinci' toy), the 2.4 g 'robot humming bird' [24], and the 3.07 g DelFly Micro (carrying a camera and transmitter) [7]. At these scales the control of sensorless brushless motors driving highly non-constant aerodynamics loads while the rotational inertia of the motor itself is incredibily small is beyond the reach of current motor controllers. Hence, for such small FWMAVs, brushed motors are unfortunately still the only option.

2.8   Conclusions

In summary, the general design concept and goal of the flapping wing MAV are the major drivers behind flapping wing design. Subsequently, choices concerning tail and wing configuration largely determine the complexity and capabilities of the design. Just as for the design of larger aircraft, basic choices have to be made regarding the type and mass of the energy source and the payload taken on board. In contrast to larger aircraft, the choices are more restricted due to the small mass. In addition, we have shown that at the scale of small flapping wing MAVs, small changes in mass can have a significant impact on the flight time of the flapping wing MAV.

In the following chapters, we will delve into the specifics of the DelFly design. A major requirement that has been present throughout the project is that the DelFly should be able to perform an observation mission. This requirement has many implications. For instance, a DelFly has to have at least one camera on board. Moreover, a DelFly has to fly for at least a few minutes so that it can fly to a different location at which it needs to perform its observations. In Chap. 3, we discuss the mechanical design choices, including the choice for the X-wing configuration. Subsequently, in Chap. 4, we explain the electronic components on board the DelFly.

References

1. Osaka slow fliers club. http://blog.goo.ne.jp/flappingwing
2. S. Avadhanula, R.J. Wood, E. Steltz, J. Yan, R.S. Fearing, Lift force improvements for the micromechanical flying insect, in *IEEE International Conference on Intelligent Robots and Systems*, 28-30 Oct 2003, Las Vegas NV (2003)
3. W. Bejgerowski, A. Ananthanarayanan, D. Mueller, S.K. Gupta, Integrated product and process design for a flapping wing drive-mechanism. ASME J. Mech. Design **131** (2009)
4. O. Chanute, *Progress in Flying Machines* (Dover, 1894, reprinted 1998)
5. Toki Corporation. http://www.toki.co.jp/

6. DARPA, The nano hummingbird surveillance and reconnaissance aircraft developed by aerovironment, inc. under contract to the united states government's defense advanced research projects agency. http://commons.wikimedia.org/wiki/File:Nano_Hummingbird.jpg (2011)
7. G.C.H.E. de Croon, K.M.E. de Clerq, R. Ruijsink, B. Remes, C. de Wagter, Design, aerodynamics, and vision-based control of the delfly. Int. J. Micro Air Veh. 1(2), 71–97 (2009)
8. G.C.H.E. de Croon, M.A. Groen, C. De Wagter, B.D.W. Remes, R. Ruijsink, B.W. van Oudheusden, Design, aerodynamics, and autonomy of the delfly. Bioinspir. Biomimet. 7(2) (2012)
9. X. Deng, L. Schenato, S.S. Sastry, Flapping flight for biomimetic robotic insects: part ii-flight control design. IEEE Trans. Robot. 22(4), 789–803 (2006)
10. X. Deng, L. Schenato, W.C. Wu, S.S. Sastry, Flapping flight for biomimetic robotic insects: part i-system modeling. IEEE Trans. Robot. 22(4), 776–788 (2006)
11. R.S. Fearing, K.H. Chiang, M. Dickinson, D.L. Pick, M. Sitti, J. Yan, Wing transmission for a micromechanical flying insect, in *IEEE International Conference on Robotics and Automation, April, 2000* (2000)
12. S.B. Fuller, M. Karpelson, A. Censi, K.Y. Ma, R.J. Wood, Controlling free flight of a robotic fly using an onboard vision sensor inspired by insect ocelli. J. R. Soc. Interface 11(97) (2014)
13. N. Gaissert, R. Mugrauer, G. Mugrauer, A. Jebens, K. Jebens, E.M. Knubben, Inventing a micro aerial vehicle inspired by the mechanics of dragonfly flight, in *Towards Autonomous Robotic Systems*, pp. 90–100. Springer (2014)
14. J. Gerdes, A. Holness, A. Perez-Rosado, L. Roberts, A. Greisinger, E. Barnett, J. Kempny, D. Lingam, C.-H. Yeh, A. Bruck Hugh et al., Robo raven: a flapping-wing air vehicle with highly compliant and independently controlled wings. Soft Robot. 1(4), 275–288 (2014)
15. J.W. Gerdes, S.K. Gupta, S. Wilkerson, A review of bird-inspired flapping wing miniature air vehicle designs. J. Mech. Robot. 4(2) (2012)
16. R. Hainsworth, L. Wolf, Hummingbird feeding. *Wildbird Magazine* (1993)
17. C-K. Hsu, J. Evans, S. Vytla, P.G. Huang, Development of flapping wing micro air vehicles - design, CFD, experiment and actual flight, in *48th AIAA Aerospace Sciences Meeting Including the New Horizons Forum and Aerospace Exposition, Orlando, Florida* (2010)
18. M. Karasek, A. Hua, Y. Nan, M. Lalami, A. Preumont, Pitch and roll control mechanism for a hovering flapping wing MAV, in *IMAV 2014: International Micro Air Vehicle Conference and Competition 2014*, Delft, The Netherlands, 12–15 Aug 2014
19. M. Karásek, A. Preumont, Flapping flight stability in hover: a comparison of various aerodynamic models. Int. J. Micro Air Veh. 4(3), 203–226 (2012)
20. M. Keennon, K. Klingebiel, H. Won, A. Andriukov, Development of the nano hummingbird: a tailless flapping wing micro air vehicle, in *50th AIAA Aerospace Science Meeting*, pp. 6–12 (2012)
21. Hobby King. http://www.hobbyking.com/
22. M. Kovac, M. Bendana, R. Krishnan, J. Burton, M. Smith, R.J. Wood, Multi-stage micro rockets for robotic butterflies. Robot. Syst. Sci. (2012)
23. N. Leichty, Micro flier radio. http://microflierradio.com/
24. H. Liu, X. Wang, T. Nakata, K. Yoshida, Aerodynamics and flight stability of a prototype flapping Micro Air Vehicle, in *2012 ICME International Conference on Complex Medical Engineering (CME)*, pp. 657–662 (2012)
25. K.Y. Ma, P. Chirarattananon, S.B. Fuller, R.J. Wood, Controlled flight of a biologically inspired, insect-scale robot. Science 340(6132), 603–607 (2013)
26. P. Muren, The 'hummer', a 1-gram flapping wing micro air vehicle, presented at EMAV 2007 (2007)
27. Plantraco. http://www.plantraco.com/
28. T.N. Pornsin-Sirirak, Y.-C. Tai, C.-M. Ho, M. Keennon, Microbat: A palm-sized electrically powered ornithopter, in *NASA/JPL Workshop on Biomorphic Robots*, Pasadena, USA (2001)
29. C. Richter, H. Lipson, Untethered hovering flapping flight of a 3d-printed mechanical insect. Artif. Life 17, 73–86 (2011)

30. P.C.S. Fuller, E. Helbling, R. Wood, Using a gyroscope to stabilize the attitude of a fly-sized hovering robot, in *International Micro Air Vechicle Competition and Conference 2014*, pp. 102–109, Delft, The Netherlands (August 2014)
31. E. Steltz, S. Avadhanula, R.S. Fearing, High lift force with 275 hz wing beat in MFI, in *IEEE International Conference on Intelligent Robots and Systems* (2007)
32. E. Steltz, R.S. Fearing, Dynamometer power output measurements of piezoelectric actuators, in *IEEE International Conference on Intelligent Robots and Systems* (2007)
33. M. Sun, Y. Xiong, Dynamic flight stability of a hovering bumblebee. J. Exp. Biol. **208**(3), 447–459 (2005)
34. G.K. Taylor, L.R.T. Adrian, Dynamic flight stability in the desert locust schistocerca gregaria. J. Exp. Biol. **206**(16), 2803–2829 (2003)
35. New Scale Technologies. http://www.newscaletech.com/
36. T. van Wageningen, Design analysis for a small scale hydrogen peroxide powered engine for a flapping wing mechanism micro air vehicle. Master's thesis, Delft University of Technology (2012)
37. C. De Wagter, The delfly micro ia a 10 cm wing span 3.07 grams flapping wing mav equipped with a camera. it was first built in 2008. https://en.wikipedia.org/wiki/DelFly#/media/File:DelFly_Micro_2008_V1.jpg (2008)
38. R.J. Wood, The first takeoff of a biologically-inspired at-scale robotic insect. IEEE Trans. Robot. **24**(2), 341–347 (2008)
39. R.J. Wood, S. Avadhanula, R.S. Fearing, microrobotics using composite materials: the micromechanical flying insect thorax, in *IEEE International Conference on Robotics and Automation 2003, Taipei, Taiwan*, pp. 1842–1849 (2003)
40. Atomic Workshop. http://www.atomicworkshop.co.uk/
41. P. Zdunich, D. Bilyk, M. MacMaster, D. Loewen, J. DeLaurier, R. Kornbluh, T. Low, S. Stanford, D. Holeman, Development and testing of the mentor flapping-wing micro air vehicle. J. Aircr. **44**(5), 1701–1711 (2007)

Mechanical Design and Materials

3

Abstract

In this chapter, we first explain the choice for the X-wing configuration of the DelFly. Subsequently, we discuss the evolution of the different mechanical parts used in the DelFly I, II, Micro, and Explorer. Details are given on the used materials.

3.1 Introduction

In this chapter, we first motivate the general concept of the DelFly design: a biplane wing model (Sect. 3.2). Subsequently, we discuss the crank mechanism (Sect. 3.3), the wings (Sect. 3.4), the tail (Sect. 3.5), and the fuselage (Sect. 3.6). For all these parts, we explain how and why they have evolved over time, as a result of measurements and flight experiences.

3.2 General Concept

The main aim of the DelFly design was to achieve an airborne camera platform with good flight characteristics. This ruled out drastic design revolutions compared to existing ornithopter designs. For example, (semi-) rigid wings articulated with many degrees of freedom were not considered in 2005. At that time, there was a lack of knowledge both on how its movements should be and could be actuated in a sensible manner.[1]

---

[1]Recent works successfully address these issues though, e.g., [7,8].

© Springer Science+Bussiness Media Dordrecht 2016

G.C.H.E. de Croon et al., *The DelFly*, DOI 10.1007/978-94-017-9208-0_3

Average flight speed	2.35 m/s	1.40 m/s	1.36 m/s
Power consumtion	0.75W	0.69W	1.00 W
Rocking amplitude	~80 mm	~0 mm	~0 mm

Fig. 3.1 Test results on model planes used to define the DelFly concept, data from [6]

Consequently, three existing ornithopter concepts were studied for the creation of the DelFly I. The first concept was a simple monoplane with one set of wings. The second was a biplane concept where two sets of wings were placed above each other. These wings moved in counter phase on a common rotational axle. The third concept involved a tandem configuration, where two wings were placed one behind the other. The wings of this concept also moved in counter phase. We wanted to test the three concepts with respect to the flight speed, the power consumption, and the stability of the ornithopter's body during flight. This last characteristic is vital for the ornithopter's role as a camera platform.

To test the concepts, models were made of simple balsa wood and commonly available tissue. The models were powered by rubber bands. The models are well known in the indoor free flight model world and were acquired as standard kits. All three models had the same span, chords, and weight. As a consequence, the wing loading of the monoplane was two times as high as the others. Each model was flight tested, while measuring the flying distance, the flight time, and the number of winds of the rubber band before and after the flight. The latter two quantities can be used to calculate the flapping frequency and the energy stored in the rubber band before and after the flight. Figure 3.1 shows the average flight speed, power consumption, and rocking amplitude for the three models.

Figure 3.1 shows the test results on model planes used to define the DelFly concept [6]. The table shows that the monoplane had the highest flight speed: a logical result of the high wing loading. For the given weight and size it consumed more power than the biplane model. Further analysis indicated that a monoplane with the same total wing area as the biplane would consume less power. Yet, this would increase the size of the MAV, contrasting the emphasis on small MAV-dimensions. The lower power consumption for the same size led us to select the biplane model. An additional advantage of the biplane model is that the low rocking amplitude of the fuselage in flight makes it a more suitable camera platform.

3.3 Crank Mechanism

DelFly I had a crank mechanism as the main wing axle, see inset I of Fig. 3.2. Its gear axles were placed in the flying direction. The advantage of this setup was that it yielded simple connecting rods with all movements in one plane. The disadvantage

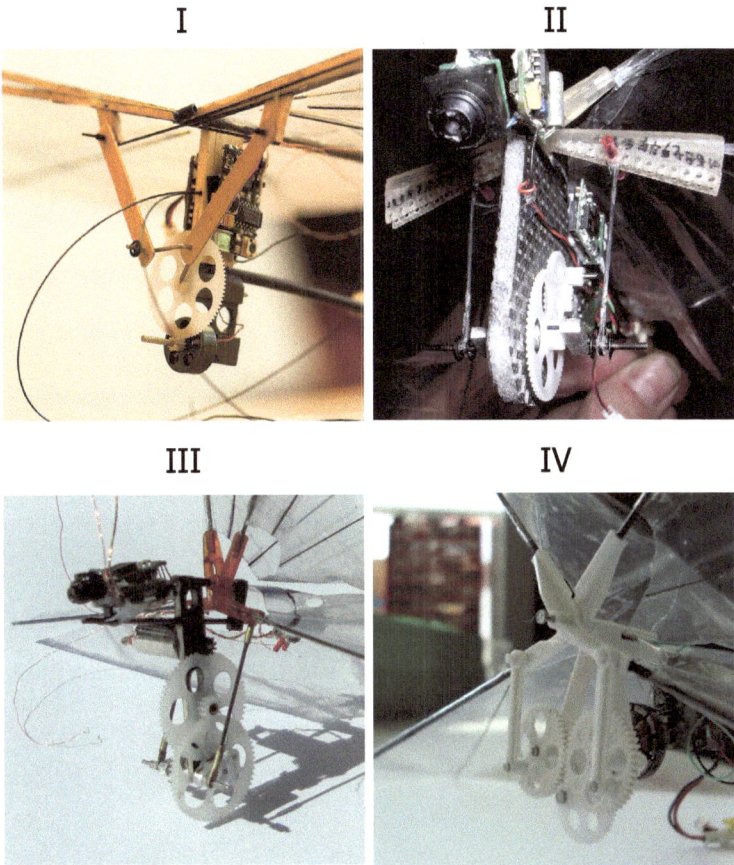

Fig. 3.2 The crank mechanism of DelFly I (*I*), the early DelFly II (*II*), the DelFly Micro (*III*), and the new DelFly II (*IV*)

was that the phase between the two sets of wings was not equal, which led to a rotational reaction movement on the fuselage and thus the camera. This was clearly visible in the camera images.

To overcome this lack of symmetry, a second drive mechanism was designed with the gear axle perpendicular to the flying direction. This setup is shown as inset II of Fig. 3.2. Used on the DelFly II, this mechanism has proven to give a good symmetrical movement with no rocking motion of the fuselage. The highest thrust was achieved with a maximum flap angle, requiring the attachment of the connecting rod as close to the fuselage as possible. A very similar design was used for the DelFly Micro (inset III in Fig. 3.2).

Fig. 3.3 3D design of the DelFly II mechanism—for more details, see [2]

The second mechanism achieved a symmetrical motion with push rods with two degrees of freedom. The manual construction of the push rods was rather complicated, with small differences from DelFly to DelFly. These small differences led to varying results in the aerodynamic experiments (see Chap. 5). In order to facilitate the production process and get more reliable measurement results, a third type of crank mechanism was designed, as shown in inset IV of Fig. 3.2. All the relevant parts, ranging from the cogs to the gear holder and push rods, have been produced via injection molding (see Fig. 3.3 for a 3D drawing).

3.4 Wings

The design of the wings started with an elliptical form that has been known for decades and is used by standard model kits. The material consisted of 6 microns Mylar foil. A stroboscope box was constructed to visualize the form of these wings during motion. The time lapse study indicated that the flexibility of such wings was too high. Inspection of the wing deformations showed that the trailing edge of the wing was completely folding during the stroke. As such, it did not contribute efficiently to the generation of lift. Therefore, we introduced stiffeners into the wing of the DelFly I. Several concepts were tested and two stiffeners gave the best result.

After flying for an extended time with the DelFly I, the foil deteriorated and a higher flapping frequency was needed for hovering. In addition, the parts of the wing surpassing the straight lines in between the end points of the stiffeners would curl up. This was an indication that these wing parts were still spurious, and they disappeared in the design of both the DelFly II and DelFly Micro.

Fig. 3.4 Wings of the
DelFly. The old wing design
of the DelFly II (*inset I*), the
new wing design of the
DelFly II (*inset II*), and the
wings of the DelFly Explorer
(*inset III*)

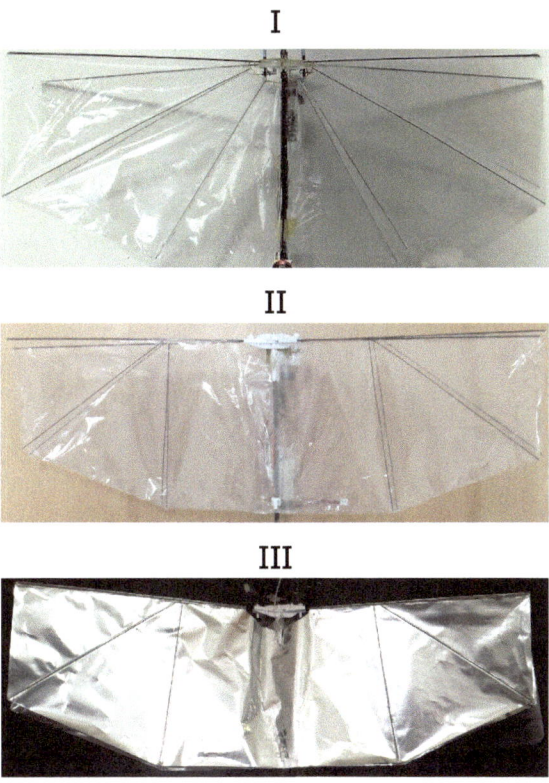

The shape of the wing is an important determining factor for the lift and the
power efficiency. Therefore, thorough studies on the relation between wing shape
and lift may contribute significantly to further minimization of ornithopters (e.g.,
[9–11]). An example of such a study is [9]. Interestingly, the shape of the DelFly II
wings (see inset I in Fig. 3.4) is very similar to their wing type 'A', which produces
a high lift at 0 wind velocity. Our own wing setup was more thoroughly tested in
later phases with high speed camera visualization [1]. Subsequently, PIV and force
measurements have been performed [5], leading to the new DelFly II wing shape,
shown in Fig. 3.4 inset II. For the DelFly Explorer, the wing design is the same, but
a different, aluminized, mylar foil is used for the wing membrane, which is slightly
less flexible and provides more thrust (Fig. 3.4 inset III).

A surprisingly important design choice for the wings has been the use of a
D-shaped rod as the leading edge. It is less stiff in the direction of the wing, which
leads to bending during the flapping cycle. This bending causes the wings to make
an 8-shape during the flapping cycle, and generates significantly extra lift: it was one
of the determining modifications that allowed the DelFly II to hover.

3.5 Tail

The DelFly I had a V-tail, because of its mechanical simplicity. An inverted V-tail was used in order to give a favorable yaw-roll coupling, making the turns without ailerons smoother. This tail can be seen in Fig. 3.5, inset I.

During the DelFly II development the inverse V-tail was abandoned for a conventional cross-tail (see inset II and III in Fig. 3.5). The reason behind that is the placement of the tail in the wake of the flapping wings during hover. With the original inverse V-tail the control of the longitudinal motion became rather marginal in this hover situation. In addition, the tail was placed closer to the wings. This

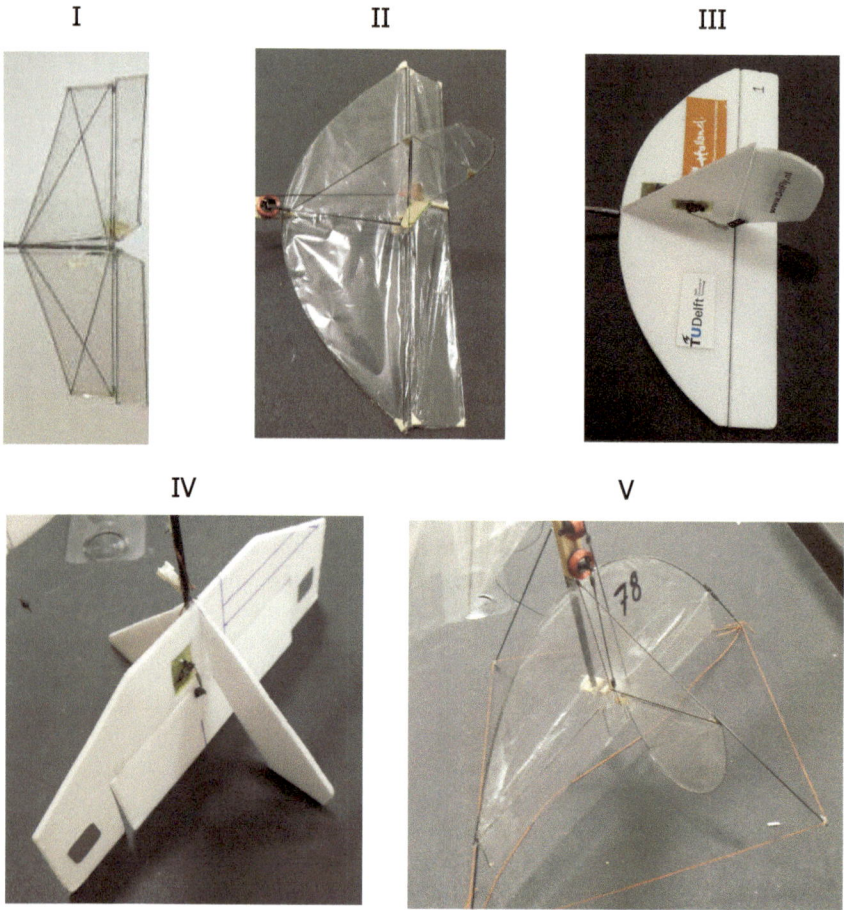

I   II   III

IV   V

Fig. 3.5 Tails of the DelFly. The inverted V-tail of the DelFly I (*I*), the old foil-tail of the DelFly II with elevator and rudder (*II*), the new foam-tail of the DelFly II (*III*). In the course of the DelFly project, tails have also been designed to support take-off and landing. Here we show the foam-tail of the DelFly Explorer (*IV*) and the foil-tail of the DelFly II (*V*)

Fig. 3.6 DelFly Explorer with ailerons instead of rudder (*bottom right*), flying toward the DelFly I model (*top left*). Reprinted with permission

improved the stability and control during the hover without reducing these aspects during cruise. The interaction of the horizontal tail with the vortices shedding from the wing might be playing a part, but this has not yet been studied.

The tail of some versions of the DelFly II is extended, such that the vehicle can take off and land as a tail sitter (see inset IV and V in Fig. 3.5). During the testing of the concept a flight has been performed with 39 take-off and landings on a single battery charge.[2]

Recently, the tail has been modified, removing the actuation of the rudder (inset IV in Fig. 3.5). In order to make the DelFly turn, two ailerons are added just behind the wings (see Fig. 3.6). The motivation behind this change is to make smoother turns. In the old rudder-design, the heading of the DelFly would change only as a second-order effect: the rudder introduces a body yaw rotation, which consequently results in a heading change. With the new aileron-design, the heading of the DelFly is influenced more directly. Namely, an aileron deflection results in a body roll rotation, which with a pitch attitude close to 90° almost directly matches the heading change.

Moreover, where the old DelFly II design uses PET-foil also in the tail, newer designs rely on foam also for the tail and ailerons. The main motivation for this is ease and repeatability of construction. The foam is less fragile than the PET-foil designs, but is also slightly heavier.

3.6 Fuselage

DelFly I and the initial designs of DelFly II used a sandwich of carbon and balsa and a self constructed carbon tube as fuselage (see Fig. 3.7). Concerning the light carbon tube, at the start of the project all commercial tubes were found to be too heavy. Hence, a tube was constructed in-house. Basically it consisted of a carbon sleeve

---

[2]https://www.youtube.com/watch?v=6OuaBWM1L-U.

Fig. 3.7 The fuselage front part and the rear end tube

that was impregnated with epoxy. When the nominal diameter of the woven carbon sleeve is small, the fibre orientation is strongly tangential which yields a high torsion stability. When the nominal diameter is large, the orientation is more longitudinal giving more bending stability. The latter is what was used on the fuselage.

The current fuselage consists of the gear box shown in Fig. 3.3 and a square 2×2 mm hollow protruded carbon tube, which is commonly commercially available. The design of the gear holder has been made such, that the carbon tube slides right in.

3.7 Conclusions

In this chapter, we have discussed the design choices and evolution of the DelFly over time. Perhaps the most essential development has been the re-design of the crank mechanism, which has significantly improved the (repeatability of the) performance. However, seemingly small choices such as that for a D-shaped rod as the leading edge can lead to significant improvements in the flight properties—in this case the lift. Another seemingly small improvement is the addition of ailerons behind the wings. Nonetheless, this addition allows the DelFly to change heading without yawing. As a consequence, the computer vision during turns is simplified. This has proven invaluable for the autonomous flight capabilities of the DelFly (see Chap. 10).

Until now, most modifications have been based on intuition, trial-and-error, but also on many measurements (aerodynamic and other). On the long term, it is desirable to have more knowledge of the aerodynamics of flapping wing MAVs and the involved structure-fluid interactions. This may allow an automated design procedure.

Acknowledgments This chapter discusses the mechanical design of various versions of the DelFly, and as such is partly based on the publications [2–6].

References

1. N.L. Bradshaw, D. Lentink, Aerodynamic and structural dynamic indentification of a flapping wing micro air vehicle, in *26th AIAA Applied Aerodynamics Conference, Honolulu, Hawaii* (2008)
2. B. Bruggeman, Improving flight performance of delfly ii in hover by improving wing design and driving mechanism. Master's thesis, Faculty of Aerospace Engineering, TU Delft, The Netherlands, 2010
3. G.C.H.E. de Croon, K.M.E. de Clerq, R. Ruijsink, B. Remes, C. de Wagter, Design, aerodynamics, and vision-based control of the delfly. Int. J. Micro Air Veh. **1**(2), 71–97 (2009)
4. G.C.H.E. de Croon, M.A. Groen, C. De Wagter, B.D.W. Remes, R. Ruijsink, B.W. van Oudheusden, Design, aerodynamics, and autonomy of the delfly. Bioinspiration Biomimetics **7**(2), 025003 (2012)
5. M.A. Groen, B. Bruggeman, B.D.W. Remes, R. Ruijsink, B.W. van Oudheusden, H. Bijl, Improving flight performance of the flapping wing mav delfly ii, in *International Micro Air Vehicle conference, Braunschweig, Germany (2010)* (2010)
6. S.R. Jongerius, M.H. Straathof, G.J. van der Veen, W.V.J. Roos, P. Moelans, R.C.A. Lagarde, A.N.A. Kacgor, C.J.G. Heynze, A. Ashok, K.M.E. de Clercq, D.A.J. van Ginneken, Design of a flapping wing vision-based micro-uav. Design synthesis exercise, Faculty of Aerospace Engineering, TU Delft, 2005
7. M. Karasek, A. Hua, Y. Nan, M. Lalami, A. Preumont. Pitch and roll control mechanism for a hovering flapping wing mav, in *IMAV 2014: International Micro Air Vehicle Conference and Competition 2014, Delft, The Netherlands, August 12–15*, 2014
8. M. Keennon, K. Klingebiel, H. Won, A. Andriukov. Development of the nano hummingbird: a tailless flapping wing micro air vehicle, in *50th AIAA Aerospace Science Meeting*, (2012), pp. 6–12
9. V. Malolan, M. Dineshkumar, V. Baskar, Design and development of flapping wing micro air vehicle, in *42nd AIAA Aerospace Sciences Meeting and Exhibit, 5–8 January, Reno, Nevada* (2004)
10. T.N. Pornsin-Sirirak, Y.-C. Tai, C.-M. Ho, M. Keennon, *Microbat: a palm-sized electrically powered ornithopter*, in *NASA/JPL workshop on Biomorphic Robots, Pasadena, USA* (2001)
11. C. Richter, H. Lipson, Untethered hovering flapping flight of a 3d-printed mechanical insect. Artif. Life **17**, 73–86 (2011)

Electronics

4

Abstract

A DelFly should be controllable, carry a payload such as a camera, and in the long term fly autonomously. This clearly requires a significant amount of electronic systems, both for payload and control. This chapter discusses all electronic components and what adaptations had to be made to employ them for light-weight flapping wing MAVs.

4.1 Introduction

The defining property of a DelFly is that it is controllable and can carry a camera. This property ensures that the FWMAV's purpose goes beyond a pure interest in flapping wing flight. An onboard camera with transmitter permits the study of the DelFly's use for observation missions. A human pilot can tele-operate the DelFly in a First Person View (FPV) mode, exploring unknown spaces.

In this chapter, we describe the various electronic components of the DelFly: the power source (Sect. 4.2), the motor (Sect. 4.3), the radio control system (Sect. 4.4), the video system (Sect. 4.6), and finally the electronics involved in onboard sensor processing (Sect. 4.7). For various subsystems there is no combination of off-the-shelf components that are sufficiently small or high performance. We explain how we improved some of the components over time for optimizing their performance on our flapping wing platform.

4.2 Power Source

The ornithopter's power source specification is an important design aspect. Its power-
and energy density is crucial for the performance. As the absolute weight of the
power source is very low many technologies that can be used in big scale are not
feasible in the scale of a Micro Air Vehicle, let alone a Nano Air Vehicle (NAV). As
discussed in Chap. 2, fuel cells, internal combustion engines, and many other power
sources are not (yet) applicable at this scale. We have not found commercial primary
batteries at the small scale of the DelFly that have a high enough power density. The
choice for Lithium Polymer batteries that has been made at the start of the project
was occasionally reviewed in the course of time but a possibly better alternative
like Lithium Sulphur batteries have not been available to us in a small enough size.
Rechargeable Lithium Polymer batteries have reached performance levels that allow
successful flight scenarios. In the DelFly we use the easily available rechargeable
Lithium Polymer batteries. The latest types of these batteries can deliver an energy
density of 170 Wh/Kg and a power density of 4000 W/Kg sustained. The versions
in the mass range of 1–4 g, relevant for MAVs and NAVs are less efficient but still
attain 130 Wh/Kg respectively 2600 W/Kg.

In the DelFly II we need about 3 W to power the motor and the electronics. With a
minimum goal of 6 min flight time this will come to a theoretical capacity of at least
0.3 Wh. Due to the relatively high load on the battery some more nominal capacity is
required. A single battery of 0.5 Wh can be found with a mass of around 4 g. As very
small batteries are less efficient than bigger ones we have chosen to use one single
battery instead of a higher number of smaller batteries. The low on-board voltage of
nominally 3.7 V has some drawbacks in the sense that e.g. motor currents are higher
than with a 11.1 V power source. The advantages of the power source are partially
reduced by the lower efficiencies of some of the electronics, however the total net
efficiency is higher.

The first years of the DelFly development the only available small powerful battery
was the 140 mAh Kokam, at that time imported in cooperation with FMA (USA) by
Ruijsink Dynamic Engineering. Since some years the availability of small excellent
LiPo batteries has increased a lot. We now use Lithium Polymer batteries from
several sources, like the 130 and 220 mAh from Atomic workshop[1] and the 180 mAh
NanoTech from HobbyKing.[2]

4.3 Motor

An electric power source requires an electric drive motor. We first explain the drive
mechanism on the DelFly. Then, we describe the consequences of this choice on our
design of the motor.

---

[1]http://www.atomicworkshop.co.uk/.
[2]http://www.hobbyking.com/.

4.3.1 Drive Mechanism

In an ornithopter we need an oscillating movement of the wings, with a frequency depending on vehicle size and flight regime. A coarse flapping frequency range is then between 8 and 50 Hz.

One could employ actuators that directly produce a linear movement with a suitable force. However, this leads to the following problem. In almost all electric to mechanic conversion, the size and mass of the system are proportional to the force. Light-weight flapping wing MAVs require an actuator to be light, so the force of the actuators is typically very limited. The power is in turn proportional to the force times the desired frequency. So, in order to reach enough power for a light actuator (with a low force), we need to operate it at a high frequency. This cannot be achieved with conventional linear or reciprocating actuators.

Still the best solution yielding a favorable power density, we have found, is a high RPM electric motor plus a crank mechanism to transform the rotation into an oscillating motion. The gear set should match the RPM with the desired flapping frequency.

4.3.2 Development of the Motor

The DelFly I featured a small brushed coreless pager motor.[3] The problem of these motors is that they provided in general only a few minutes of sustained flight, and usually wore out within an hour. The brushes are too fragile for the current at the required power and the low efficiency of 35 %, led to overheating and expansion of the rotor inside the motor. This caused the motor to block itself. In general these motors have a too low power to weight ratio.

For the DelFly II we developed a brushless motor that matches the requirements in an ornithopter, in collaboration with the company DC Enterprises in India [3]. The resulting brushless motor (see Fig. 4.1) was more efficient than the motor on the DelFly I, and delivered enough mechanical power to be able to hover (1.5 W or 1 kW/kg).

A brushless motor can reach higher efficiencies and higher power densities than conventional brushed motors. In addition, there are no other wearing parts than the (ball) bearings. This makes the reliability of the motors several orders of magnitude better than their brushed counterparts.

A brushless motor has a stator with 3 electric phases and an electronic controller that produces a revolving electro-magnetic field. This makes the magnet rotor rotate. The accurate timing of the revolving magnetic field can be accomplished by the controller with the aid of magnetic sensors in the motor, which can give a well determined positional feedback of the rotor. These sensors can be avoided in more sophisticated controllers where the position of the rotor in relation to the stator

---

[3]http://www.didel.com/, type MK07-2.3, red.

Fig. 4.1 Image of a
brushless motor

phase is measured by means of the electromotive force (EMF) voltage from the non-energized winding. The timing can be determined rather easy in this way when the motor runs smoothly with a constant or only slowly changing RPM.

In the development of the DelFly motor and controller we have found that in our application the determination of the timing is hampered by two aspects of uneven rotation.

1. The first prototype of the motor did run very well with a propeller but would not run well with only a small polyacetal pinion on the axle. The low rotational inertia of the rotor and pinion (lack of flywheel effect), lead to a fluctuation of the speed of the motor within each rotation.
2. Secondly there is the uneven load during one flapping cycle. Near the point where the flapping direction changes, the load on the motor changes abruptly also leading to timing difficulties.

The first problem was related to the cogging torque of the motor. The discrete number of magnets of the rotor interacting with the stator poles produces an often significant torque fluctuation on the rotor. This causes very swift speed change within each cycle of the rotation. At high RPM the motor did run acceptably smooth and the timing was adequate, but the motor would not run at medium and lower speeds also necessary in our application. In collaboration with Dr. E. Lomonova of the Eindhoven University we have made a magnetic simulation of the motor to find ways of reducing the cogging torque. With the help of simulations, we studied whether it was possible to reduce the cogging torque by varying the magnetic structure of the motor, the number of magnets, and the magnetic embracement: the percentage covered by magnets over the circumference of the rotor. The original motor had $3 \times 3 = 9$ stator phases and 12 magnetic poles a standard setup als found in many small CD-Rom and Hard-Disk motors. It had a magnetic embracement of almost 100 %. In the simulation using the programm Maxwell-2D, the embracement was varied gradually by changing the sizes of the magnets in the motor. Figure 4.2 shows the simulation result of the relation between the embracement (in %, x-axis) and the efficiency (in

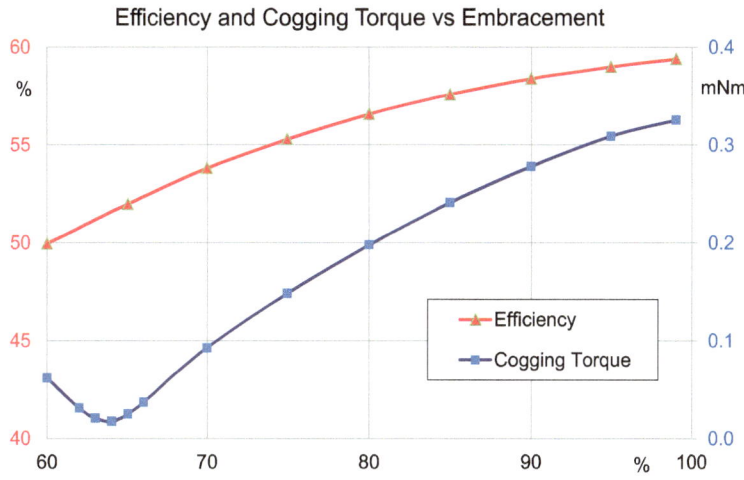

Fig. 4.2 Effect of the embracement on the efficiency and the cogging torque

%, left y-axis) and cogging torque (in N*mm, right y-axis). It showed that reducing the embracement would reduce the cogging torque. A minimum cogging torque was realized with an embracement of 64 % a significant decrease from 330 to 17 Nmm. However, the efficiency of the motor would also decrease from 59 to 52 %, which was not what we were aiming for.

Another option to reduce cogging torque was a slightly bigger change in the magnetic structure of the brushless motor. Brushless motors can be made with a vast range of phase-pole combinations, some excellent, some marginal. In general the number of stator phases is always a multiple of 3 where they have an even number of magnetic poles. In the smaller electric motors the 9 phase-12 pole and 12 phase-14 pole approach are the most common but others can behave well. A 9 phase-10 pole version was simulated and found to be very suitable.

The simulations showed that with the 10 magnets an embracement of 83 % would lead to practically zero cogging torque with an efficiency as high as the original motor. This could be achieved with 10 of the original 12 magnets spaced evenly around the circumference of the rotor, with an embracement of ≈83 %. Without new tooling, an extremely low cogging torque was promised with an equally high efficiency as the base version (while maintaining the same RPM of the electromagnetic field). Due to the reduced number of magnets the actual mechanical RPM was 20 % higher (see Fig. 4.3). This formed no problem, as a two stage gearing was needed anyway and only the ratio had to change. In addition, the winding scheme for a 10 pole motor is different from the 12 pole version but that was easily implemented. Finally, since 2 magnets were removed the final motor was also lighter. A brushless motor with 9 phases and 12 Poles is wound with all windings in the same direction and spread around the cicumfence of the stator. We normally refer to the phases as A, B and C, in the case of a 9ph/12pole motor the winding scheme is written as A-B-C-A-B-

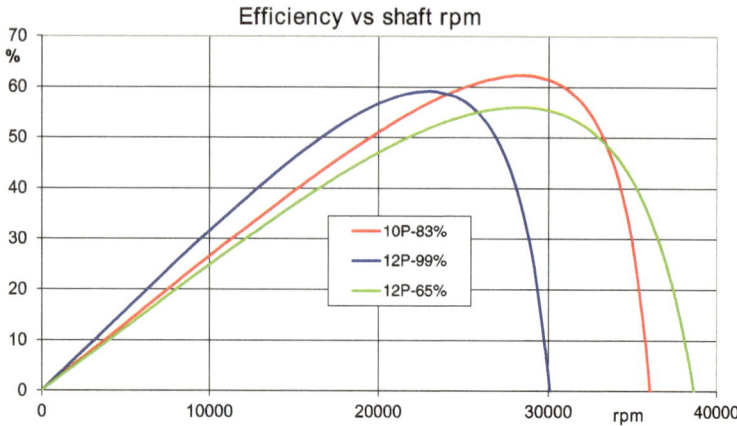

Fig. 4.3 Efficiency versus shaft RPM for different brushless motor configurations

C-A-B-C. This winding scheme does not work for our 9ph/10pole motor, here we need another winding scheme with also windings in the reverse direction. This is indicated with lower case letters. We now need a winding scheme that can be written as: A-a-A-B-b-B-C-c-C. The resulting motor performed very much as simulated, this design was used in all our further work.

The second timing problem was caused by the uneven load during the flapping cycle of the ornithopter. The RPM changes twice in the range of a factor of two during each flapping cycle. When the wings absorb their maximum power halfway the stroke, the motor RPM drops to almost half the RPM value at stroke reversal where the wings absorbs very little power. On top of that, even the slightest wear or dead play in the gear and pushrod systems leads to shocks when going from pulling to pushing on the wings. These quick load changes are problematic for the commutation of the motor. In order to run smoothly and efficiently the motor controller has to do the commutation some 20° before zero crossing of the EMF signals. This can only be done by estimating the current rotation speed and predict when the commutation will take place. When the motor speeds up too quickly or too slowly this may lead to *misfiring*. To solve this problem, we cooperated with both Jean Louis Coural of Micro Plane Solutions (France) and MicroInvent (Slovakia). The solution involved both the hardware and the software of the motor controller. The EMF voltages needed to be measured more accurately by the hardware, and the measurements needed to be filtered differently by the software algorithms. In particular, there was a larger emphasis on measurements with respect to the predictions. The resulting controller is shown in the left part of Fig. 4.4.

To match the maximum required flapping frequency with the RPM range of the electric motor a gear ratio of 20:1 has been adopted. The gears are from polyacetal and have a module of 0.3. We have a 12 teeth pinion on the motor a 48/12 teeth idler gear. On the 60 teeth final gear a dual crank is fitted to actuate the two sets of wings (see right part of Fig. 4.4).

Fig. 4.4 *Left* Motor controller designed with Micro Plane Solutions of France, in order to cope with the uneven load during one flapping cycle. *Right* The gears as used on the DelFly II

4.4 Radio Control System

A solid control of an MAV is equally important as a low weight. For the DelFly I, we have used the first version of a commercial 900 MHz system by Plantraco [2]. This resulted in good control, but also in many small problems. These problems included:

- a way too low PWM frequency to power the then used brushed pager motor,
- no current limit on the actuator outputs made it not enough foolproof,
- awkward frequency control,
- and low quality transmitter sticks.

Still, the system had as main advantage that it was very light.

Consequently, when designing the DelFly II, we searched for a different kind of system. When the [11] Minor receivers came on the market we used them in conjunction with our standard 35 MHz transmitters. These receivers are very versatile and can be used with all kind of actuator and motor types.

Around 2008 DelFly II changed to a modified 2nd version Plantraco system. The Martin Newell *Rabbit* HipHop system, operates still on 868 or 900 MHz, but now with different soft- and hardware. It is light to extremely light, and without all the problems we have encountered before. The system uses a frequency hopping algorithm to avoid any problem that could arise from frequency clashes, when flying in halls or during shows. The transmitter used is a HF module that couples to a decent transmitter front-end. This is also the technology used in the DelFly Micro.

In 2011 a new 2.4 GHz receiver was used developed by DelTang [12]. The RX31 is only about 9.3×9.9 mm and weighs only 0.21 g. It uses DSM2 technology and as such is compatible with all Spectrum transmitters. The new DelFly models are all equipped with this receiver (Fig. 4.5).

Fig. 4.5 DelTang 2.4 GHz
0.21 g DSM2 remote control

4.5 Actuators

The rudder and elevator of the DelFly are typically actuated by magnetic actuators
(see Fig. 4.6). At the beginning of the DelFly project these were the only systems
of under 1 g to provide adequate control. Conventional servos were available with a
mass of around 2 g at that time.

Currently there are more options emerging, such as light-weight muscle-wire or
Piezo actuators. Muscle-wire is a Shape Memory Alloy (SMA), which can become
smaller by applying heat to it. Although it is in principle promising for light-weight
actuation of ornithopters, its response time is currently too long for optimal control
and the efficiency is too low. Piezo actuators are more efficient, but still have too
high voltage requirements for the batteries on board the DelFly.

The employed magnetic actuators offer just enough control power, at a reasonably
low weight and power consumption. They provide a smooth and fully proportional
control. The actuators we use are the Plantraco MiniACT [2].

Fig. 4.6 Magnetic actuators
on the DelFly II

In the meantime, 'conventional' servos have also been miniaturized further. There-fore, on the newer DelFly models we also use light-weight servos, such as the 0.45 g servo from Microflierradio or even the 1.1 g servo from Hobbyking as described in Chap. 2. These are equally light and fast, but more powerful and accurate.

4.6 Video System

One of the principles of the DelFly is that it is always equipped with a camera system. The camera on the DelFly I and early versions of DelFly II has a dimension of 8 × 8 × 7 mm (see Fig. 4.7). It is a special version of the MO-S588 1/4 in. CMOS NTSC camera with 380 lines and a 3.1 mm lens. This color version has a sensitivity of 1.5 Lux, while the EIA Black and White version has a sensitivity of 0.05 Lux. The camera needs a clean power supply of 5 V at 40 mA. We developed a switched capacitor DC-DC converter in-house, because all inductor types that we have tried gave interference problems either to the camera or to the R/C system. Our inductorless version is based on the LTC3200. It is mounted directly on the rear of the camera. The camera with DC-DC converter weighs exactly 1 g. The transmitter works down to 3.3 V and can be driven directly from the flight battery. When in 2009 more sensitive cameras and wider field of views came on the market, DelFly II was equipped with those.

As will be further explained in Chap. 10, in 2012 our attention turned to stereo vision systems for autonomous flight. To this end, two cameras were placed on board together with an image transmitter (see Fig. 4.8). The stereo camera system consisted of two synchronized CMOS 720 × 480 cameras (with an offset of 7.6 cm) running at 25 Hz, and a 2.4 GHz NTSC analog transmitter. The cameras have a field of view of 68° horizontally and 50° vertically.

Fig. 4.7 Two pictures of the DelFly I/II camera. *Left Side view* of the camera. *Right* Modified *back-side* of the camera

Fig. 4.8 Stereo vision
cameras and image
transmitter for autonomous
flight experiments

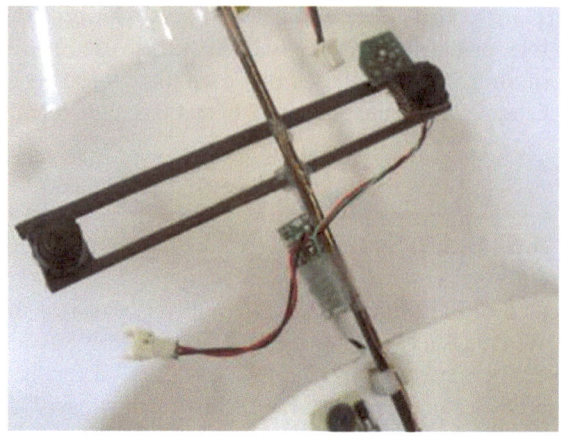

Because there is only one transmitter, the video streams from both cameras have to be combined as one. The merging of the images can be performed in various manners, with different consequences for the stereo vision process. In CMOS cameras, an image is obtained by reading all pixel values one after another. NTSC video frames consist of a field of even image lines and a field of odd image lines. To combine the video streams from the two cameras in one transmitted image, a switch selects which camera output is used as the source for a certain pixel. The standard way of combining the images is to select the first camera to read all pixels on the even lines and then to switch to the other camera to read the odd image lines. This results in two images that both have a resolution of 720×240 pixels.

The issue with this way of image merging is the time delay between the two images. At a frame rate of 25 Hz, it takes roughly 40 ms to read all pixels from the CMOS sensor. This means that it takes 20 ms to read all pixels from the even lines only. The time delay between the first rows of the even and odd fields is then also 20 ms. In terms of camera motion on a FWMAV, this is a significant delay. For the DelFly, the flap cycle period is typically around 70 ms.

A solution for this is to switch between the two cameras each time a full line has been finished. Instead of switching after reading 240 lines, switching is done after reading each line. The delay between image lines is then reduced by a factor of 1/240 to roughly 83 µs. The image lines from the left and right image first fill the even lines, and only later the odd lines. As a result of putting a left and a right image on the even lines the resolution of the images is further reduced to 720×120 pixels. In total 480 scan lines are still used, but only 120 lines of the left camera and 120 lines of the right camera belong to the even field. The same holds for the image lines that belong to the odd field, but the fields cannot be combined because of the 20 ms delay.

4.7   Onboard Sensor Processing

In the introduction it was mentioned that the DelFly must be able to carry a camera, so that it is useful for tele-operation. However, tele-operation has its limits. One major limitation stems from the limited connection range of wireless image transmission. When entering a building with the DelFly II, the human operator will be able to explore the first few rooms, but the video link will be lost when there are multiple walls in between the FWMAV and the operator. In addition, there may be applications, such as surveillance of large industrial sites, where there is a need for many FWMAVs flying in the same time.

In such cases, it is desirable that the FWMAV can fly completely by itself. In order to achieve this, it is not sufficient to carry the sensors on board. The FWMAV will also have to carry a processor that can interpret the sensor data and transform it into control actions. This section discusses different onboard processing elements important to the DelFly's autonomous flight capabilities.

4.7.1   Autopilot

In 2011 DelFly II was equipped with its first autopilot: the *WiiPilot1* (Fig. 4.9). An *ATMega88pa* in MLF28 package formed its core and measured 4 by 4 mm. This fast and efficient 8 bit microcontroller could read a dual Invensense Gyro, namely the IDG500. The 2.4 GHz receiver from DelTang [12] can be configured to communicate over a single wire. This was then fed to the micro-controller and stabilization signals from the gyroscopes could be overlaid on the remote control signals.

A heading hold system was added. The autopilot was aligned with one gyro sensor parallel to gravity. The fact that DelFly makes turns with almost no roll angle in slow hovering flight allowed a simplistic single gyro heading hold to be created. Since no external sensors were available the heading hold was constructed by negatively integrating the heading change commands in the gyro integration. Gyro bias was calibrated on startup. The PID controller would always steer the heading integral to zero. All computations are done in fixed point. To reduce the flapping vibrations a first order lowpass was used in the damping term.

$$\Psi_{i+1} = \Psi_i - RC_{command} + (\omega_z - bias_z) \tag{4.1}$$

Besides gyroscopes *WiiPilot1* could also be equipped with a WiiMote camera. When stripped to the bare chip, this camera with onboard processing weighs only 0.4 g. This system was able to track up to 4 infrared LED. This system was also used in [1] to control an FWMAV for following an LED target. On the DelFly II we have used this system to achieve precision flight inside a wind tunnel [7, 10].

Some DelFly II were equipped with an SCP1000 barometer. This chip can provide barometric pressure in high resolution mode at 5 Hz. Thanks to the nice control properties of DelFly II, this enabled the construction of an altitude hold mode using

Fig. 4.9 WiiPilot1

barometer only. The following equations show the altitude control as was first shown in [6]:

$$h_{pascal} = p_{baro} - p_0 \tag{4.2}$$

$$h_{LP} = h_{LP} + (h_{pascal} * 64 - h_{LP})/64; \tag{4.3}$$

$$h_{LIM} = limit(h_{LP}, -40, 50) \tag{4.4}$$

$$\delta_{Thrust} = PID h_{LIM}, \tag{4.5}$$

where p_0 is the pressure at the ground level (measured at the start of the flight), p_{baro} the current measured pressure, h_{pascal} the relative pressure, and h_{LP} a low-passed version of h_{pascal}. A limited version of h_{LP} with as minimum -40 and maximum 50 is used as the error in a PID altitude controller.

4.7.2 Vision Processing

In 2013 we have introduced the DelFly Explorer, the first FWMAV that can autonomously explore unknown environments [8]. The main facilitator for this autonomous flight was a new onboard stereo vision system, shown in Fig. 4.10.

Fig. 4.10 4 g stereo vision system—reprinted with permission

The stereo vision system has two digital cameras with a baseline of 6.0 cm and an STM32F405 processor. Importantly, the flapping motion of FWMAVs introduces deformations in the camera images [5,9]. Therefore, it is not possible to use subsequently recorded left and right images for stereo matching [13]. The cameras of the stereo system are synchronized and provide *YUYV* image streams, and in the current implementation a Complex Programmable Logic Device (CPLD) merges the streams from both cameras by alternately taking the *Y* component of the stream from both cameras. This results in a single image stream with the order $Y_l Y_r Y_l Y_r$. The resulting stream contains simultaneously sampled pixels at full camera resolution but without color.

4.8 Conclusions

It is well-known that the mobile phone market has spurred the developments in miniature electronics. Major drivers for the mobile phone market, such as low weight and power, correspond with the needs of a light-weight flapping wing MAV. Still, this chapter has shown that many electronic components used in mobile phones needed some modification before being incorporated into a flapping wing MAV. This for instance to further reduce the weight or to deal with FWMAV peculiarities such as the flapping wing motion and related vibrations. Toward the future, we expect that the miniaturization will continue, e.g., leading to small commercially off-the-shelf onboard vision systems. We hope that this chapter helps to quickly adapt newly developed electronic components to flapping wing MAVs.

Acknowledgments This work would not have been possible without the help and expertise of many. Bob Selman Designs (USA), DC Enterprises (India), Micro Plane Solutions (France), MicroInvent (Slovakia), Martin Newell (USA), MicroFlier-Nick Leichty (USA), DelTang (UK). The chapter is partly based on [4,8,13].

References

1. S.S. Baek, F. Garcia Bermudez, R. Fearing, Flight control for target seeking by 13 gram ornithopter, in *IEEE/RSJ International Conference on Intelligent Robots and Systems* (2011)
2. Bob Sellman. http://www.bsd.com/
3. U. Chandrashekhar. Mighty midget micro brushless motors. http://microbrushless.com/
4. G.C.H.E. de Croon, K.M.E. de Clerq, R. Ruijsink, B. Remes, C. de Wagter, Design, aerodynamics, and vision-based control of the Delfly. Int. J. Micro Air Veh. **1**(2), 71–97 (2009)
5. G.C.H.E. de Croon, E. de Weerdt, C. de Wagter, B.D.W. Remes, R. Ruijsink, The appearance variation cue for obstacle avoidance. IEEE Trans. Robot. **28**(2), 529–534 (2012)
6. G.C.H.E. de Croon, M.A. Groen, C. De Wagter, B.D.W. Remes, R. Ruijsink, B.W. van Oudheusden, Design, aerodynamics, and autonomy of the Delfly. Bioinspir. Biomim. **7**(2), 025003 (2012)
7. C. De Wagter, A. Koopmans, G.C.H.E. de Croon, B.D.W. Remes, R. Ruijsink, Autonomous wind tunnel free-flight of a flapping wing MAV, in *EuroGNC*. Delft (2013)
8. C. De Wagter, S. Tijmons, B.D.W. Remes, G.C.H.E. de Croon, Autonomous flight of a 20 gram flapping wing MAV with a 4 gram onboard stereo vision system, in *2014 IEEE International Conference on Robotics and Automation (ICRA)* (2014)
9. F. Garcia Bermudez, R. Fearing. Optical flow on a flapping wing robot, in *IROS* (2009), pp. 5027–5032
10. J.A. Koopmans. Delfly freeflight—autonomous flight of the delfly in the wind tunnel using low-cost sensors. Master's thesis, Delft University of Technology, 2012
11. MicroInvent. http://www.microinvent.com/
12. D. Thenissen. DT rc control systems. http://www.deltang.co.uk/
13. S. Tijmons, G.C.H.E. de Croon, B.D.W. Remes, C. De Wagter, R. Ruijsink, E-J. Van Kampen, Q. Chu, Stereo vision based obstacle avoidance on flapping wing mavs, in *EuroGNC* (2013)

Part II
Aerodynamics

Introduction to Fixed and Flapping Wing Aerodynamics

5

Abstract

In this chapter, physical principles of fixed-wing and flapping-wing aerodynamics are introduced briefly with the associated terminology. Furthermore, the main aerodynamic mechanisms which are responsible for the generation of the forces in the flapping-wing flight are described in order to provide the reader with fundamental knowledge before further comprehensive analysis of the DelFly.

5.1 Introduction

Man-kind has been inspired by flapping-wing flight of birds and insects for ages. However, as discussed in Chap. 1, the first flight of humans happened in a balloon rather than with wings in the late 18th century. It took more than another century until the first powered airplane flight with fixed wings was performed thanks to the diligent efforts of Wright brothers, particularly between 1899 and 1905 [27]. Since then, our knowledge about fixed-wing aerodynamics has increased substantially, whereas mysteries of flapping-wing aerodynamics remained mostly unresolved. However, the ever-increasing interest in the field of Micro Air Vehicles (MAVs) has redirected attention to flapping wing aerodynamics particularly in the last two decades as conventional modes of flight (i.e., fixed and rotary wing) are relatively insufficient in terms of efficiency and maneuverability in the intended scales.

The aerodynamics of fixed-wing flight has been investigated thoroughly in the last century and is mostly well-understood from low-speed to high-speed flow regimes [32]. Although it is not in the scope of this book to elaborate on the topic of fixed-wing aerodynamics, a short introduction of the force generation mechanism for the case of a wing in a steady motion is given in Sect. 5.2 with the associated terminology.

The aerodynamic principles, however, as applied in fixed-wing aerodynamics, are not sufficient to explain generation of relatively high forces in flapping-wing insect flight as discussed in the famous 'bumblebee paradox'. It was shown by Demoll

© Springer Science+Bussiness Media Dordrecht 2016   57
G.C.H.E. de Croon et al., *The DelFly*, DOI 10.1007/978-94-017-9208-0_5

[11] that once principles of steady attached flow aerodynamics are applied to the flight of bumblebees by taking into account also the speed of the beating wings, they would need a lift coefficient over twice that of any aircraft [8]. Relatively recently, this paradox has been solved by showing that insects exploit unsteady aerodynamic mechanisms to produce such large forces and thereby stay aloft [7, 14, 20]. In Sect. 5.3, these mechanisms are explained in a general sense as well as in relation to the flapping-wing mechanism of the DelFly.

5.2 Fixed Wing Aerodynamics

In this section, first the generation of circulatory lift is described in the most basic aerodynamic situation: a two-dimensional airfoil in a low-speed incompressible flow starting from rest. Then, forces and flow around three-dimensional finite-span fixed wings are discussed briefly. Note that comprehensive analysis of fixed wing aerodynamics is out of the scope of this book and the reader is referred to aerodynamics and fluid dynamics textbooks written by [3, 32, 33, 35].

However before starting the discussion of fixed-wing aerodynamics, it will be useful for the reader to give a short explanation of some wing geometry definitions and general aerodynamic terms. Figure 5.1 depicts a wing planform which can be found in a general aviation aircraft. The front and back edges of the wing are called the leading edge and the trailing edge, respectively. The distance between these edges is the chord length which is denoted by the symbol c. Right and left ends of the wing are called the wing tips and the distance between the wing tips is the span length (b). In this wing planform, the chord length varies over the span such that the wing narrows toward the tip ($c_{root} > c_{tip}$). This type of wing planform is called the tapered wing planform and widely preferred in aviation industry due to its aerodynamic and structural advantages. In the calculation and normalization of the aerodynamic terms, the mean chord length ($\bar{c}$) is used as the characteristic length of the wing, which is calculated by dividing the wing projected area (S) by the span

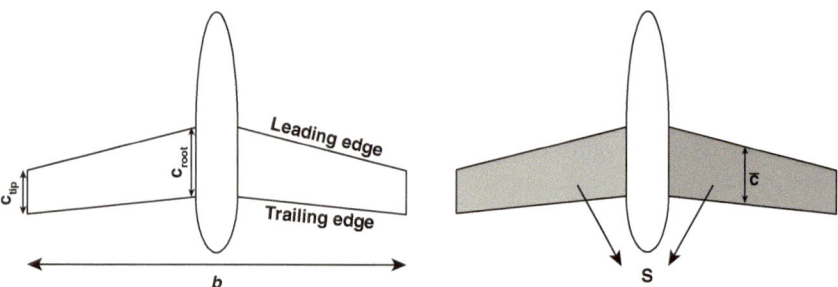

Fig. 5.1 Geometrical definitions of a wing planform

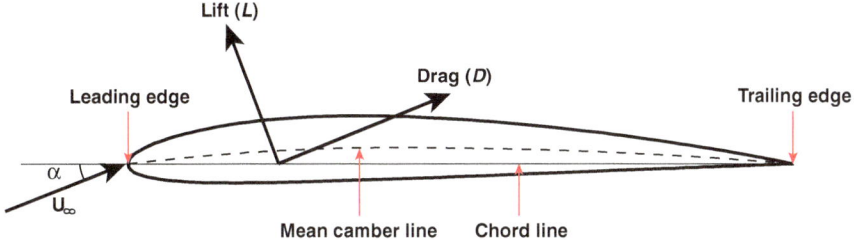

Fig. 5.2 A cambered airfoil placed in a free-stream flow

length. Another important geometrical definition is the aspect ratio (AR), which is calculated as follows:

$$AR = \frac{b^2}{S} = \frac{b}{c} \tag{5.1}$$

Aspect ratio is a measure of the slenderness of a wing and for a rectangular wing planform, it is simply the ratio of the span length to the chord length. Aspect ratio is an important parameter in terms of generation of aerodynamic forces which will be discussed in the following sections. The chordwise sectional cut of the wing is the wing profile that is also known as the airfoil. As an example, a cambered airfoil with a round leading edge and a sharp trailing edge is shown in Fig. 5.2. The line connecting the leading and the trailing edges is the chord line and the curve that is halfway between the upper and lower contours of the airfoil is the mean camber line. For the representation of the aerodynamic forces, the airfoil is positioned in a free-stream with a velocity of U_∞. The angle between the chord line and the free-stream vector is the angle of attack (α). The aerodynamic force component that is perpendicular to the oncoming flow direction is lift (L) and the parallel component is drag (D). In the terminology of aerodynamics, however, more fundamental quantities are used to express the aerodynamic forces, which are dimensionless force coefficients. To calculate lift (c_L) and drag (c_D) coefficients, forces are non-dimensionalized by the dynamic pressure ($q_\infty = 1/2\rho U_\infty^2$ where ρ is the fluid density) times the wing projected area (S). Henceforth, in order to denote the force coefficients of a three-dimensional wing, capital subscript letters will be used (c_L, c_D), whereas small subscript letters (c_l, c_d) will be used in order to indicate coefficients for a two-dimensional airfoil such that they are calculated per unit span.

5.2.1 Generation of Circulatory Lift in Two Dimensional Airfoils

There are various explanations about the generation of lift in the popular scientific literature. One of these explanations is based on the variation of pressure around the airfoil. Due to the shape of the airfoil, the velocity of the fluid molecules displays local variations. For an airfoil moving in air or placed in a free-stream flow in a lift generating configuration, the air molecules move faster over the top surface of the airfoil than those over the bottom surface. As can be inferred from the Bernoulli's equation, the variation of velocity around the airfoil surface also results in varying pressure values such that relatively low pressure region is formed over the upper surface with respect to the lower surface. This formation results in the generation of the aerodynamic force with its component perpendicular to the flow direction being lift. Another explanation is primarily based on the Newton's third law of motion such that an airfoil in flow directs the moving fluid particles downward so that it experiences an opposite reaction aerodynamic force (i.e., lift). However in this section, we focus on a more fundamental explanation of generation of circulation and lift around an airfoil.

The lift per unit span for an airfoil placed in a free-stream is given by the famous Kutta-Joukowski theorem as:

$$L' = \rho \times U_\infty \times \Gamma \qquad (5.2)$$

where ρ is the density of fluid, U_∞ is the free-stream velocity and Γ is the circulation around the airfoil. At this point, circulation appears as an important quantity which is defined as the line integral of velocity around any closed surface that encapsulates the airfoil.

To find out how the flow develops such a circulation, let us consider an airfoil at an angle of attack of α which is set in motion from initial state of rest. Lift is not generated instantaneously and the flow around the airfoil is similar to the potential flow pattern (irrotational and incompressible) as shown in Fig. 5.3a. In this early stage of the flow development, the stagnation point is positioned on the rear upper

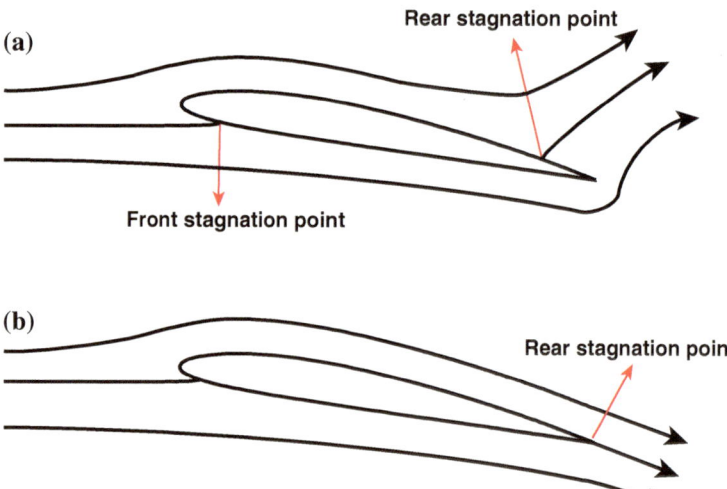

Fig. 5.3 Potential flow solutions around an airfoil: **a** flow with zero circulation ($\Gamma = 0$); **b** flow with a finite circulation ($\Gamma = \Gamma_{bound}$) that establishes the Kutta condition at the trailing edge

surface of the airfoil and the flow tries to curl around the sharp trailing edge from the bottom surface to the top surface. Theoretically, the flow velocity becomes infinite at the sharp corner under inviscid (neglecting effects of viscosity) and irrotational flow conditions, causing also formation of infinitely large velocity gradients. In reality, however, this is not physically possible and even a slightest viscosity with these very high velocity gradients will result in the generation of high viscous forces around the trailing edge. At this stage, the flow is unable to turn around the sharp corner because of viscosity and a region of intense vorticity emanates from the trailing edge and rolls up into a vortex, which is called the starting vortex. In consequence, the rear stagnation point moves toward the trailing edge until the steady flow condition is reached. In essence, this is similar to adding a circulatory flow component to the initial flow field, which will increase the flow velocity over the upper surface and decrease it along the lower, thereby moving the rear stagnation point towards the trailing edge. For a unique value of circulation (Γ_{bound}), the stagnation point is placed exactly at the trailing edge. In this case, the flow is leaving the upper and lower surfaces of the airfoil smoothly at the trailing edge and tangential to it (Fig. 5.3b). This phenomenon is called the "Kutta condition" and once satisfied it ensures that the airfoil at an angle of attack imparts a downward momentum on the fluid. which results in the generation of an upward reaction force (i.e., lift) acting on the airfoil.

According to Kelvin's circulation theorem, the circulation around a closed curve that encloses the same fluid elements remains constant, if the circuit remains in a region of inviscid flow (note that viscous processes can still occur in the region enclosed by the curve). Now, consider a circuit that is large enough to include the initial and final positions of the airfoil as well as the same fluid particles. As the circuit does not intersect with the boundary layer, where viscous effects are not negligible, but positioned far away from the airfoil, the inviscid flow assumption is plausible. At the beginning of the motion, the circulation is zero around this closed curve. As the motion of the airfoil starts, the starting vortex with a counterclockwise circulation sheds into the wake. In view of Kelvin's circulation theorem, the flow around the airfoil should have equal and opposite circulation to that of the starting vortex. This circulation is known as the bound circulation and its value is also the unique value that moves the rear stagnation point exactly to the trailing edge and generates lift particular to the specific condition of the airfoil. However, when the angle of attack or the free-stream velocity changes, the circulation around the airfoil changes as well to adjust the flow to establish the Kutta condition and a new starting vortex is shed from the trailing edge.

Lift for a given angle of attack can be calculated for a thin airfoil in the context of thin airfoil theory. The airfoil is so thin that its effect on the flow can be simulated by a vortex sheet distributed along the chord line. The distribution is such that it establishes the Kutta condition at the trailing edge (i.e., the strength of the vortex sheet at the trailing edge is zero). This particular distribution is then integrated from the leading edge to the trailing edge to calculate the unique bound circulation value which is then used in Eq. 5.2 to calculate the lift. As a result of this theoretical approach, the lift coefficient is found to be linearly proportional to the angle of attack with a slope of 2π.

Theoretical calculation of the lift coefficient based on the thin airfoil theory and under inviscid flow conditions displays a good agreement with the experimental lift data up to a certain angle of attack. Although theoretically the lift coefficient increases with increasing angle of attack, in reality the airfoil experiences a sharp decrease in lift after a certain degree. This phenomenon is known as the *stall* of the airfoil and it occurs due to flow separation at angles of attack higher than a critical angle particular for each airfoil (*stall angle*). The flow separation is a viscous phenomenon and cannot be predicted by the inviscid theory. that explains why the above theoretical approach yields continuously increasing lift coefficient with increasing angle of attack. However, in real flow conditions where viscous effects are present, the flow over the top surface cannot resist the adverse pressure gradient and the effect of friction on the top surface of the airfoil at high angles of attack which causes the flow to separate from the surface resulting in the stalling of the airfoil. Stall characteristics and hence maximum lift coefficient depends on the airfoil geometry (in particular the thickness of the airfoil) and the Reynolds number [1].

5.2.2   Aerodynamic Characteristics of Finite Span Wings

In the previous section, we explained the force generation mechanism and aerody-namic properties of an airfoil, which can be essentially considered as a wing with an infinite span. It might be counter-intuitive for the reader to see that aerodynamic properties of a three-dimensional wing are different from those of an airfoil as it is simply a section of the wing. However, three-dimensional character of a finite-span wing results in a three-dimensional flow over the wing. To explain how this is formed, let us consider a finite span wing placed in a free-stream flow and generating lift. Thus, physically there exists a pressure difference between the bottom and top surfaces of the wing: high pressure at the bottom surface and low pressure at the top surface. As a result of this pressure imbalance, the flow tends to curl around the wing tips in the direction from the high pressure region to the low pressure region. This circulatory motion gives rise to the generation of tip vortices (Fig. 5.4) and spanwise flow pattern on both sides of the wing such that it is directed from root to tip at the bottom surface and from tip to root at the top surface.

An important aerodynamic effect of this circulatory flow at the wing tips is the generation of downwash (w), which is a small velocity component in the downward direction induced by the tip vortices. This downward velocity component together with the free-stream velocity produce a local relative stream that is inclined with respect to the free-stream direction (see Fig. 5.4). The angle of inclination is called the induced angle of attack (α_{ind}). As a result, the local wing profile experiences a smaller angle of attack than the geometrical angle of attack, called the effective angle of attack ($\alpha_{eff} = \alpha - \alpha_{ind}$). In turn, there exists now a component of the lift vector in the direction of the free-stream, called the induced drag (D_{ind}). This drag

Fig. 5.4 Tip vortex, downwash, induced angle of attack and formation of induced drag for an airfoil placed in a free-stream flow

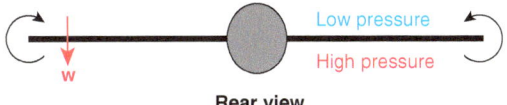

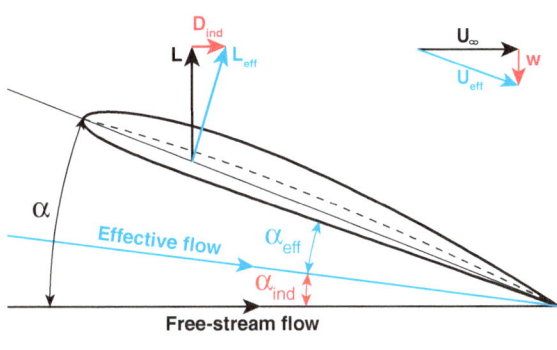

component adds to the profile drag component which derives from the skin friction drag (D_f) and pressure drag (D_p), both of which are apparent in viscous flows [32].

The first theory to calculate the aerodynamic properties of a finite-span wing was developed by Prandtl, namely the lifting line theory. In this theory, he represented the wing with an infinite number of horseshoe vortices which are superimposed over the lifting line and calculated the generation of downwash due to this formation. Then he finally assessed the circulation distribution over the wing span. One of the most important results of his theory is about the induced drag coefficient ($c_{D,ind}$), which reads:

$$c_{D,ind} = \frac{c_L^2}{\pi e AR} \tag{5.3}$$

where e is the span efficiency that varies in the range of 0-1 and gets the maximum value of 1 for an elliptical wing [32]. It is evident that an elliptical wing, which also has an elliptical lift distribution in regard to the Prandtl's lifting line theory, yields the minimum induced drag. Furthermore, higher aspect ratio wings generate lower induced drag accordingly. Obviously, generation of lift increases the induced drag, which is not surprising as the same pressure imbalance that creates lift also causes the formation of induced drag. On the other hand, the lift slope ($a = dc_L/d\alpha$) for a general three-dimensional wing is given as:

$$a = \frac{a_0}{1 + (a_0/\pi AR)(1 + \tau)} \tag{5.4}$$

where a_0 is the lift slope for an infinite span wing. The lift slope of a three-dimensional wing is proportional to the aspect ratio meaning that high aspect ratio wings generate higher lift at a given angle of attack.

In conclusion, it is clear that aerodynamic properties of a finite span wing differ from those of an airfoil due to three-dimensional nature of the flow field. In this context, wing aspect ratio and spanwise chord length distribution emerge as effective geometrical parameters on the generation of forces for a three-dimensional wing.

5.3   Flapping Wing Aerodynamics

In this section, flapping wing motion, as observed in natural fliers, is first discussed by describing different motion kinematics and associated terms. Subsequently, force generation mechanisms of flapping-wing flight are addressed.

5.3.1   Kinematics of Flapping Flight

The natural flapping flight of birds and insects is a complicated three-dimensional phenomenon which arises from combined motions of the flapping wings and the flyer's body. In order to stay aloft or maneuver, the natural flyers drive their wings in three main motions which are sweeping, pitching and plunging (heaving).

In general, the wing motion during flapping strokes can be considered to be confined in a plane, which is referred to as the stroke plane (see Fig. 5.5). For simplicity in the current description of flapping wing kinematics, the stroke plane is considered to be horizontal but it might also be inclined with respect to the horizontal axis with an angle of β. It was shown that biological flyers adjust the orientation of the stroke

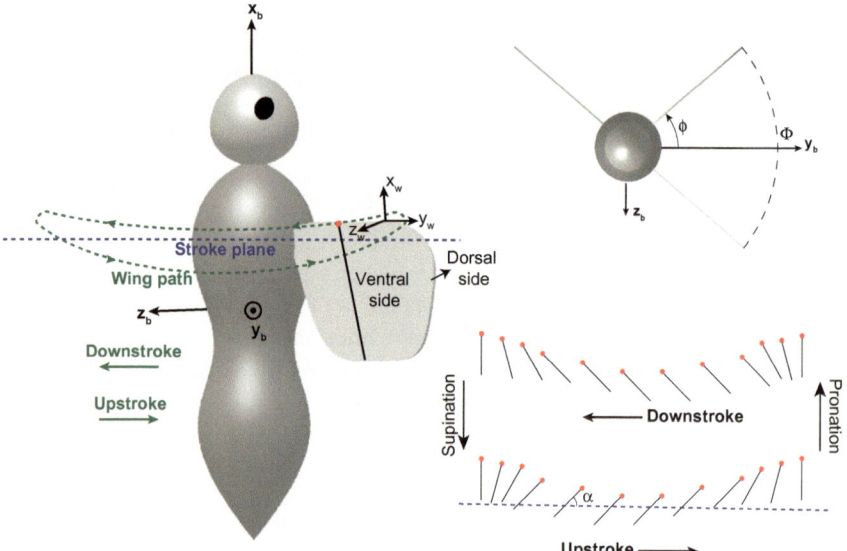

Fig. 5.5 Flapping wing motion in biological flyers

plane actively in an effort to maneuver as it affects the force generation and direction of the resultant force vector [60]. The fore and aft motion of the wing in the stroke plane is the sweeping motion. The part of the motion in which the ventral side of the wing leads is the downstroke and that with the dorsal side leading is the upstroke. The stroke angle (ϕ) is defined as the instantaneous angle between the leading edge of the wing and the y axis in the body coordinate system (y_b), whereas the stroke amplitude (Φ) is the total angular displacement of the wing in the stroke plane.

The pitching motion is the rotational motion of the wing about an axis parallel to the y axis in the wing coordinate system (y_w). It is generally performed during stroke reversals in order to have a positive angle of attack (α) in the subsequent stroke. The wing pitch reversal at the transition from an upstroke to a downstroke is called pronation, and at the transition from a downstroke to an upstroke, it is called supination. The pitching motion can contribute to the force production significantly during the flapping flight depending on a number of parameters such as its timing with respect to the sweeping motion of the wing and the pitch rate. These aspects of the motion are discussed in detail in Sect. 5.3.2.2.

The up and down motion of the wing with respect to the mean stroke plane is the heaving (plunging) motion. The plunging motion plays an important role in determining wing tip trajectory pattern, which has a major influence on the production of aerodynamic forces [49]. Many studies that are focused on the flapping-wing motion in insects and hummingbirds revealed two main patterns: a figure-of-eight pattern, in which the wing crosses through the stroke plane once during each stroke and once during the stroke reversal (total of four times plane-crossing in a complete cycle); an oval trajectory in which the plane-crossing occurs between the stroke reversals [17]. Obviously, these patterns, mostly due to the heaving motion component, determine the effective angle of attack experienced by the wing during the stroke and in turn affects the force and moment generation. For this reason, variations of the flapping patterns have been subject to a number of studies aiming to assess an optimum pattern with an efficient force generation [2,37,49].

5.3.1.1 Non-dimensional Analysis of Flapping Flight

Non-dimensional analysis is an appropriate tool for the characterization of the system under consideration and to find out which combination of parameters are important under given conditions. Moreover, non-dimensional parameters are used to define regimes in which different systems behave similarly. In this context, three main non-dimensional parameters related to the fluid dynamics in flapping flight are addressed in this section for different flight regimes [54].

The Reynolds number (Re) is the ratio of inertial forces to viscous forces and therefore quantifies the relative importance of these forces for given flow conditions. In general terms, it is defined as:

$$Re = \frac{\rho_f L_{ref} U_{ref}}{\mu} \qquad (5.5)$$

where ρ_f is the fluid density, L_{ref} is the characteristic length, U_{ref} is the reference velocity and μ is the dynamic viscosity of the fluid. Similar to fixed-wing aerodynamics, the mean chord length ($\bar{c}$) is utilized as the reference length. However, in the case of flapping flight, the selection of the reference velocity is not as straightforward and requires careful consideration of the given flight conditions. In forward flight, common practice is to use forward flight velocity (U_∞). It is also plausible to use the mean wing tip velocity ($U_{m,tip} = 2\Phi f R$, where f is the flapping frequency and R is the wing semi-span), particularly in the situation of slow forward flight and high flapping frequency. In particular, the mean wing tip velocity is commonly used as the reference velocity in hovering flight. The corresponding definition of the Reynolds number can then be rewritten as:

$$Re = \frac{\rho_f f \Phi b \bar{c}}{\mu} = \frac{\rho_f f \Phi A R \bar{c}^2}{\mu} \tag{5.6}$$

where $b = 2R$ is the wing span and AR is the aspect ratio ($AR = b^2/S$ with $S = b\bar{c}$ being the wing planform area). Alternatively, for hovering flight the Reynolds number may be defined in relation to the velocity induced by the flapping wings (V_{ind}). The induced velocity can be approximated by use of the actuator disk theory.

The actuator disk theory is originally known as the Rankine-Froude momentum theory, which is generally used to determine the induced velocities in the wake of propellers from the momentum flux required to provide a given propeller thrust [19]. In this theory, the propeller is replaced by an infinitely thin rotating disk of an area S_d (Fig. 5.6), which does not apply a resistance to the air that flows through it. Therefore, the velocities above and below the disk are assumed to be equal ($V_2 = V_3 = V_d$) yet there is a pressure jump ($P_2 \neq P_3$) as the rotating disk introduces energy to the

Fig. 5.6 Schematic representation of the actuator disk (Rankine-Froude momentum) theory

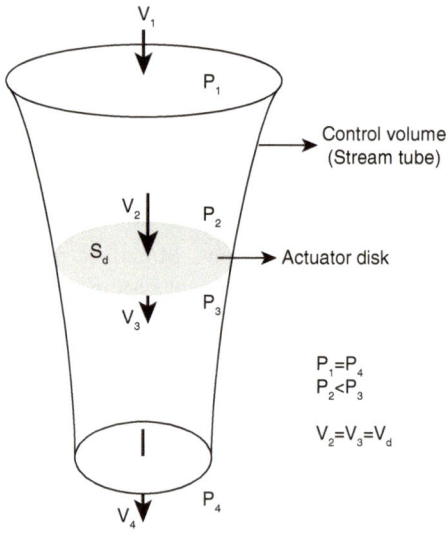

flow. The flow is assumed to be inviscid, irrotational and steady. Moreover thrust loading and velocity are assumed to be uniform over the disk area. Thrust generated by the disk can be determined by use of the pressure difference between the lower and upper surface as follows:

$$T = (P_3 - P_2)S_d \tag{5.7}$$

Defining the streamtube as a control volume with upstream and downstream boundaries far enough such that the static pressure values are equal to free-stream pressure (or ambient pressure in case of hovering flight with no free-stream velocity) and applying Newton's second law, thrust can also be calculated using the momentum fluxes entering and leaving the control volume:

$$T = \dot{m}\,(V_4 - V_1) \tag{5.8}$$

where $\dot{m}$ is the mass flow rate through the streamtube ($\dot{m} = \rho S_d V_2$), V_1 is the velocity at the upstream boundary and V_4 is the exit velocity. By applying Bernoulli's equation for the regions above and below the disk separately, it is found that the velocity at the position of the disk is:

$$V_d = \frac{V_1 + V_4}{2} \tag{5.9}$$

For a hovering flight configuration, where V_1 is zero, Eq. 5.9 indicates that the induced velocity (V_{ind}) is half of the far-wake exit velocity ($V_d = V_{ind} = V_4/2$). Using this relation and thrust being equal to the weight of the flyer in hovering flight in Eq. 5.8, the induced velocity can be calculated as follows:

$$V_{ind} = \sqrt{\frac{W}{2\rho_f S_d}} \tag{5.10}$$

For propellers the active disk area can be defined as $S_d = \pi L^2$, where L is the length of the propeller blades. However, for the application of the actuator disk theory in flapping flight, it will be more convenient to define the actuator disk area by taking into account the stroke angle (Φ) and the stroke plane angle (β) [19]. Then, the disk area is $S_d = \Phi R^2 \cos \beta$, through which the flapping wings impart downward momentum.

The Strouhal number is a non-dimensional parameter that is relevant for the vortex dynamics and shedding behavior of vortices in flapping-wing aerodynamics. In the case of forward flight, with the full stroke amplitude of the flapping motion (ΦR, i.e., wake width) as the characteristic length and the forward flight velocity as the reference velocity, it reads:

$$St = \frac{f \Phi R}{U_\infty} \tag{5.11}$$

In this form, Strouhal number compares the flapping velocity of the wing with the forward flight velocity, which can be considered as a measure of propulsive efficiency for a flapping-wing flyer in forward flight [53].

The reduced frequency is an important similarity parameter that can be defined as the measure of unsteadiness in flapping-wing flight. In forward flight, based on the mean chord length, the free-stream velocity and the angular frequency of the flapping motion ($2\pi f$) it is defined as:

$$k = \frac{\pi f \bar{c}}{U_\infty} \tag{5.12}$$

The reduced frequency is essentially the ratio of mean chord length to the wavelength of shed vortices in forward flight. In hovering flight, because the forward flight velocity is zero, the mean wing tip velocity is used as the reference velocity resulting in:

$$k = \frac{\pi \bar{c}}{\Phi b} = \frac{\pi}{\Phi AR} \tag{5.13}$$

It is clear that the reduced frequency definition based on the mean wing tip velocity in hovering flight is not a function of flapping frequency but inversely proportional to the stroke amplitude and the aspect ratio, which is counter-intuitive with respect to the physical interpretation of this parameter. Therefore, it is more suitable to use the induced velocity (Eq. 5.10) as the reference velocity because it can be considered as the convection velocity of the vortical structures.

5.3.2   Force Generation Mechanisms

Natural fliers exploit numerous different unsteady mechanisms for the generation of forces during flapping flight. The extent to which these mechanisms are used varies between different species based on length scales, wing kinematics and flight regime. Moreover, these mechanisms might be used in case of necessity during a particular phase of the flight. For instance, to perform a sharp maneuver or to carry an extra weight, the flying animal can enable an aerodynamic mechanism simply by changing its wing kinematics. These unsteady mechanisms are discussed in the review papers provided by Sane [51]), Lehmann [36], Shyy et al. [54] and explained in [53,57] to a great extent. In this section, we will cover the most prominent and relevant mechanisms given as follows:

1. Leading edge vortex (delayed stall)
2. Rotational forces
3. Clap-and-fling motion
4. Wake capture
5. Added mass

5.3.2.1   Leading Edge Vortex (delayed Stall)
Contrary to fixed-wing flight, flapping-wing flight mainly relies on flow separation and the associated pressure fields for the generation of forces. In this respect, the phenomenon of the leading edge vortex (LEV) emerges as one of the most dominant mechanisms in flapping-wing aerodynamics particularly in insect flight.

   The term 'delayed stall' is used to indicate the period of time when the airfoil generates relatively high lift with the presence of an attached LEV prior to stall. Let us consider an airfoil with a sharp leading edge that is set in motion at an angle of attack above its stalling angle. The flow separates at the leading edge but may reattach before it reaches to the trailing edge such that the Kutta condition is maintained. In such a case, an LEV forms in the separation zone. The presence of the LEV creates a low pressure region on the top surface of the wing which enhances the lift. Equivalently, the extra lift can be explained as an increase of the circulation around the wing with the contribution of the LEV so that the airfoil imparts a greater momentum to the fluid [20].

   Although the discovery of the LEV in insect flight is relatively recent, lift enhancement effects of vortices have been studied for a long time [9,30,47,62]. Polhamus [47] described an analytical model to predict the force generated by the presence of a vortex, namely the *leading edge suction analogy*. Although this analogy was originally developed to assess the lift characteristics of a sharp-edge delta wing at low speeds, it was also used in order to explain the LEV phenomenon in flapping-wing flight [36]. In case of a thin airfoil placed in a free-stream with a low angle of attack, the flow is completely attached on the upper surface and the Kutta condition is established at the trailing edge. According to the thin airfoil theory, we showed that lift is generated as a result of the bound circulation of the wing in Eq. 5.2. When this force is decomposed into components in the chord-normal and -tangential directions, we end up with a positive upward normal force and a tangential force parallel to the chord line in the upstream direction (see Fig. 5.7a). The latter component is actually associated with the acceleration of the flow around the leading edge and thus the formation of a low pressure region. The two-dimensional aerodynamic drag becomes zero due to the leading edge suction phenomenon under inviscid and incompressible flow conditions, whereas in reality there is always some drag exerting on an airfoil placed in a free-stream flow due to presence of viscosity (i.e., skin friction drag). This contradiction is also known as d'Alembert's paradox [33]. However, as the angle of attack increases and the flow separates at the leading edge, the leading edge suction force vanishes due to formation of an LEV. Polhamus' analogy states that the increase in the normal force component due to the LEV related pressure distribution is equal to the suction force that is necessary to accelerate the flow around the leading edge [28]. In other words, with the increase of the angle of attack beyond the stall limits and the formation of the LEV, the leading edge suction force rotates 90° (see Fig. 5.7b) and increases the force component normal to the airfoil surface (thus increasing lift and drag).

   This mechanism provides a significant contribution to the force generation as long as there is a stable attached LEV. For a two-dimensional translational motion of a wing, the size of the LEV increases continuously until the flow reattachment is no longer possible. The LEV then sheds into the wake and new trailing edge vorticity forms at the trailing edge breaking the Kutta condition. Stalling of the wing causes a significant decrease in the force generation [51]. Dickinson and Götz [13] performed force measurements and flow visualizations on translating two-dimensional flat plate wings undergoing an impulsive start at different angles of attack. Their results showed

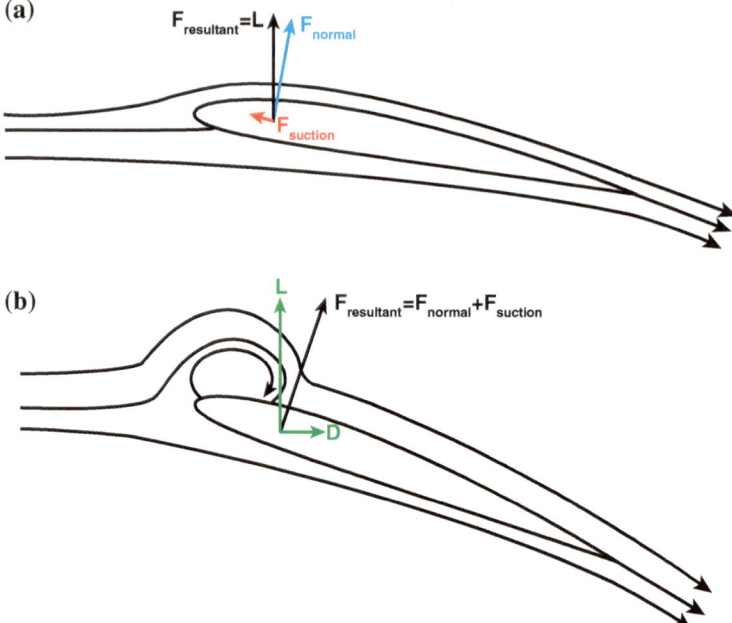

Fig. 5.7 Schematic representation of the Polhamus' leading edge suction analogy: **a** formation of low pressure region in front of the leading edge causes formation of the suction force ($F_{suction}$) and tilting of the resultant force vector toward upstream direction, **b** increasing angle of attack results in the formation of an LEV and the generation of the suction force in a direction normal to the airfoil surface

that at angles of attack above 13.5°, an LEV is formed as a result of the impulsive start and increases lift dramatically for the first two chord lengths of travel after which the LEV sheds. This is accompanied by the development of a counter-rotating trailing edge vortex (TEV). They observed that this alternating pattern of vortices leads to a von Karman vortex street (see Fig. 5.8).

The first direct evidence of the LEV mechanism in insect flight was provided by Ellington et al. [20]. They visualized the flow around the flapping wings of the hawk-moth *Manduca sexta* and observed an intense LEV that is created by the dynamic stall mechanism during the downstroke. Then in order to investigate the phenomenon, they built a three-dimensional robotic flapping device that mimics the wing kinematics of a hovering hawkmoth and carried out a flow visualization study around the flapping wings. Contrary to the shedding of the LEV in the two-dimensional motion of an airfoil, the LEV stays attached to the wing during the complete downstroke, which they explained with the presence of a spanwise flow under three-dimensional

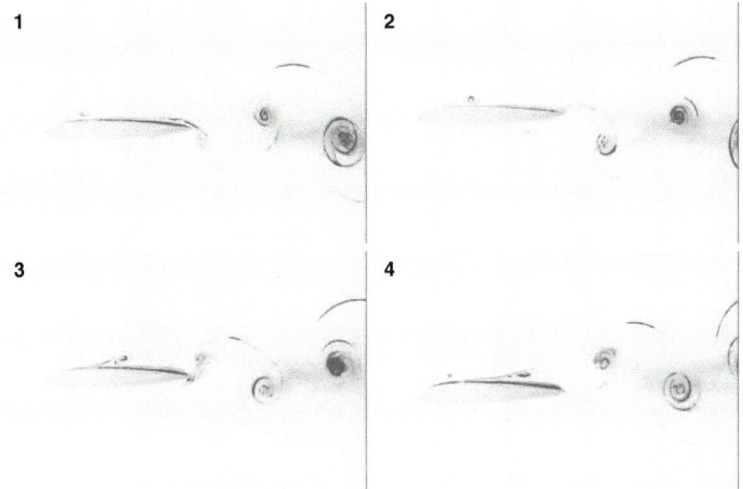

Fig. 5.8 Flow around a plunging NACA 0012 airfoil placed in a free-stream flow visualized via tin precepitation method (1. Half-way upstroke, 2. The end of upstroke, 3. Half-way downstroke, 4. The end of downstroke): The shedding of the LEV and TEV from the two-dimensional airfoil gives way to the formation of reverse von Karman vortex street (adapted from [45])

conditions. Their measurements indeed revealed that the LEV has a spiral conical shape enlarging toward the wing tip and feeding into the tip vortex after 70 % of the span. There is also a strong spanwise flow in the core of the LEV with the maximum values comparable to the wing tip velocity. Based on these observations, they hypothesized that the spanwise flow convects some of the leading edge vorticity toward the wing tip which inhibits excessive growth of the LEV analogous to the mechanism of vortex stabilization that occurs on delta wings [20].

However, flow field measurements on a dynamically scaled robotic model of a fruit fly (*Drosophila*) wing operating in hover conditions at a Reynolds number of 160 revealed that in contrast to the flapping hawkmoth wings, the axial velocity in the vortex core is marginal, at around 2–5 % of the average wing tip velocity as reported by [4]. Instead, it was found that a relatively strong base-to-tip flow pattern is present in the rear two-thirds of the wing reaching 40 % of the tip velocity. They used teardrop fences at two different chordwise positions in order to block the spanwise flow in the core of the LEV and in the rearward region of the wing in two separate experiments to test the influence of base-to-tip flow on the attachment of the LEV. The forward blockage case, rather than increasing the LEV strength and leading to a possible separation, resulted in a decrease in the vortex strength and thus in the net force. Although this finding rebuts the spanwise flow hypothesis for the vortex stability, it should still be noted that relatively low Reynolds number of the fruit fly operation might lead to significant differences in the flow structures. Their alternative hypothesis to explain the prolonged attachment of the LEV was based on

(a)   **(b)**

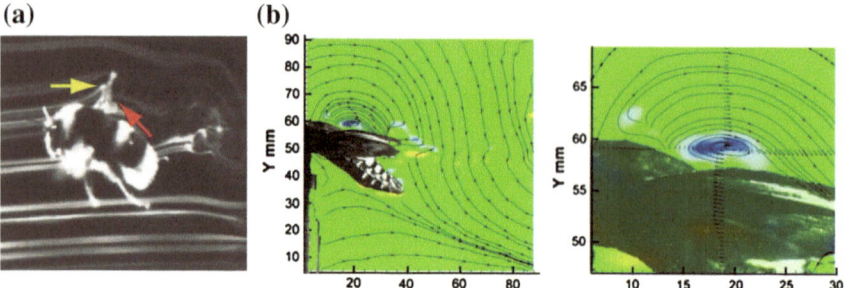

Fig. 5.9 **a** Smoke flow visualization around a bumblebee reveals formation of an LEV with the reattaching smokeline (indicated with *red arrow*) depicting the extent of the LEV (adapted from [8]; **b** flow around the wings of an hawkmoth (*Manduca sexta*)) and in its wake (visualized by the contour plots of vorticity) shows the formation of an LEV (adapted from [6]) - reprinted with the kind permission of Springer Science and Business media

the downwash generated by the tip vortex, wake vorticity and associated decrease of angle of attack. The lower effective angle of attack in turn limits the growth of the LEV. More recently, Lentink and Dickinson [39] claimed that the rotational accelerations (i.e. centripetal and Coriolis accelerations), which are effective at low Rossby numbers, are stabilizing the LEV. Clearly, there is a consensus about the attachment of the LEV in three-dimensional flapping-wing motion but the underlying stabilization mechanism still remains mainly unresolved.

The presence and benefit of the LEV phenomenon were also addressed in the three-dimensional flapping-wing model of the fruit fly (*Drosophila melanogaster*) [14]. In this study, the elevated force coefficients during the reciprocating motion of the wing are attributed to three main mechanisms: delayed stall, wing rotation and wake capture. The LEV mechanism works during the sweeping part of the motion, whereas the latter two are effective during stroke reversals. Free-flying butterflies (*Vanessa atalanta*) [56] and bumblebees (Fig. 5.9a) [8] also exploit an LEV to generate forces. The use of LEV mechanism is encountered even at larger scales, i.e. in slow-flying bats [42]. It was shown that these bats are able increase lift up to 40 % by exploiting the LEV and reaching a maximum lift coefficient of 4.8.

5.3.2.2 Rotational Forces

As mentioned previously, a reciprocating flapping-wing motion of a biological flier consists of two translational phases in which the wing sweeps at high angles of attack and during which delayed stall and LEV phenomena emerge as dominant force production mechanisms. In order to preserve a positive angle of attack during these translational phases, the wing performs pronation and supination as described in Sect. 5.3.1. Although the LEV mechanism accounts for most of the force that keeps an insect aloft, it cannot provide sufficiently high forces that an insect needs during steering maneuvers [14] or while carrying loads at the order of its body weight

[40]. In this sense, instantaneous force measurements on dynamically scaled fruit fly wings revealed considerable additional force generation during stroke reversals indicating the potential of wing rotation as an important unsteady mechanism [14]. Although the significance of pitching motion in flapping flight has been realized in the last decades, it has been already addressed extensively, particularly in the context of wing flutter, in theoretical studies [23,25,43,58] which were supported by experimental studies of [22,24,29,34,48,55] as cited in [51].

In the context of the unsteady thin airfoil theory, force production as a result of pitching motion is a circulatory phenomenon. When a wing starts rotating about a spanwise axis (viz. pitching) during its revolution or translation, the Kutta condition breaks down and the stagnation point departs from the trailing edge which leads to a formation of shear and vorticity. In order to re-establish the Kutta condition, additional circulation is generated around the wing that either adds to or subtract from the available circulation depending on the relative direction of the pitching motion. This additional circulation due to rotation of the wing is estimated by [23] using a quasi-steady equivalent-downwash model for a thin airfoil performing a rotational motion around a spanwise axis as follows [18]:

$$\Gamma_{rot} = \pi(0.75 - \hat{x}_0)\omega c^2 \qquad (5.14)$$

where ω is the rotational velocity and $\hat{x}_0$ is the non-dimensional distance of the rotation axis from the leading edge ($\hat{x}_0 = x_0/c$). It is clear that the circulation associated with the rotational motion is a function of rotational velocity and the position of the rotation axis. According to the theoretical estimate, the critical axis, at which the circulation changes sign, is at the three-quarter chord position, which is also verified experimentally by [26,50] for relatively high rotational velocities.

In addition to the position of the rotation axis, timing of the rotation plays an important role on the force generation. Force measurements on the robotic flapping fruit fly wings revealed different force histories for advanced, symmetrical and delayed rotations during stroke reversals [14,50]. In the case of the advanced rotation, in which the wing rotation precedes the stroke reversal, rotational circulation has a constructive effect on the lift. This can be interpreted as an increase of angle of attack while the wing approaches to the end of the stroke. When the rotation is delayed with respect to the stroke reversal (viz. delayed rotation), the leading edge rotates forward relative to the translating motion resulting in a negative lift during the stroke reversal. If the wing rotation is symmetrical about the stroke reversal, the wing generates first upward and then downward force. Clearly, the timing of the wing rotation can be so influential such that a phase advance of 8 % from the delayed to the symmetrical case results in a 67 % of increase in the mean lift coefficient [14].

The rotational circulation has a remarkable lift enhancing effect in flapping-wing flight such that its contribution can reach 50 % of the total lift generation averaged throughout the stroke cycle despite its short application time [14]. Furthermore, by changing the timing of the rotation, the force direction can be adjusted for instance

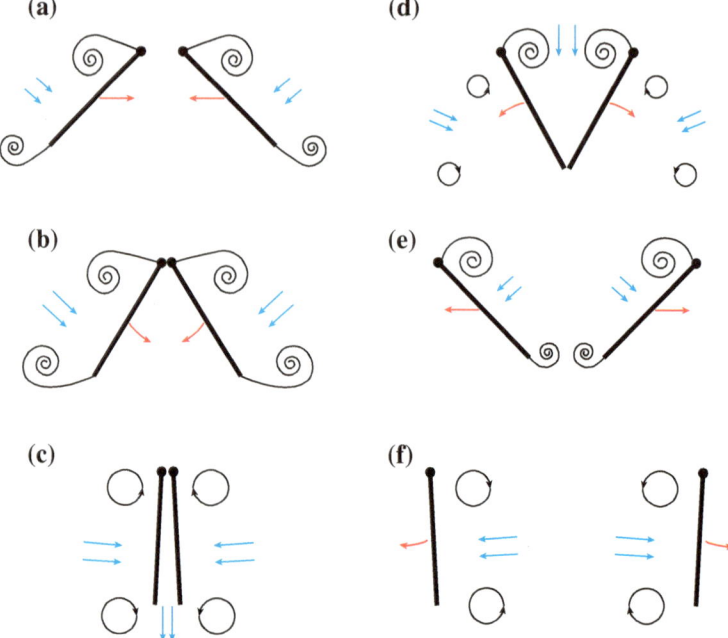

Fig. 5.10 Schematic representation of clap-and-peel motion by use of two rigid wing sections with the *red arrows* indicating the direction of the wing motion and the blue arrows indicating the direction of the induced flow, for the instroke (**a–c**) and outsroke (**d–f**)

to perform an intended maneuver. Studies on fruit flies showed that they change the wing rotation timing and stroke amplitude during their flapping flight [15].

5.3.2.3 Clap-and-fling Motion

Clap-and-fling is a lift enhancement mechanism which was first described by [61]. This relates to the wing-wing interaction phenomenon, which takes place at dorsal stroke reversal (Fig. 5.10). During the clap phase, the leading edges of the wings approach each other (Fig. 5.10a) and pronation about the leading edges occurs until the v-shaped gap between the wings is closed (b and c). In the fling phase, the wings rotate about their trailing edges forming a gap in between (d). Subsequently, the wings start to translate apart from each other (e and f). Investigations on birds and insects showed that as well as being used continuously during flight, some species utilize this mechanism for a limited time in order to generate extra lift, especially while carrying loads or during the take-off phase [36]. Experiments on insects [40] showed that use of the clap-and-fling mechanism results in a generation of 25 % more aerodynamic lift per unit flight muscle than conventional flapping-wing motions.

Several studies have attempted to provide an explanation of the increase in the force generation due to clap-and-fling motion. The underlying unsteady aerodynamic mechanisms for this increase have been proposed as: (1) the attenuation of starting vortex formation at the onset of the fling (thus diminishing the Wagner effect) due to the interaction of opposing circulation of each wing [51] (see Fig. 5.10d, e); (2) the downward momentum jet formed at the end of clap, when the air trapped between the wings is forced downwards (c); (3) generation of a massive leading edge vortex (LEV) at the onset of fling (d), which was verified experimentally by Lehmann et al. [38]. They performed simultaneous PIV and force measurements on dynamically scaled rigid fruit fly wings in order to investigate the effects of the clap-and-fling motion on the force production. They pointed out that clap-and-fling motion, depending on the stroke kinematics, may enhance the force production by up to 17 %. Detailed PIV analysis revealed that the existence of a bilateral image wing increases the circulation induced by the leading edge vortex during the early fling phase, obviously correlated with a prominent peak in both lift and drag. Furthermore, it was shown that trailing edge vorticity shed during the fling phase of the motion is considerably reduced with respect to that in the single flapping wing case.

The flexibility of the wings plays an important role in the wing kinematics and force generation mechanism of the clap-and-fling motion. As a result of the wing flexibility, the fling phase occurs more like a peel, while the clap phase can be considered as reverse-peel [17]. This is the reason why clap-and-fling is called clap-and-peel motion in case of flapping flexible wings. It was speculated that flexible wings increase lift by enhancing the circulation in the fling phase and boosting the strength of downward momentum jet in the clap phase [18]. Moreover, it was indicated that flexibility reduces drag by allowing the wing to bend or reconfigure under the aerodynamic loading. Miller and Peskin [41] investigated this phenomenon computationally by use of an immersed boundary method for a Reynolds number of 10. They found that clap-and-fling with flexible wings produces lower drag and higher lift with respect to clap-and-fling with rigid wings.

5.3.2.4   Wake Capture

At the end of each stroke in flapping motion, the wing sheds counter-rotating leading and trailing edge vortices into the wake. Subsequently as a new stroke starts after a stroke reversal, the wing encounters its own wake which contains high energy regions (see Fig. 5.10c, d). These regions are formed by the induced velocities of the wake vortices that are shed in the previous stroke. Transfer of the energetic fluid's momentum to the wing by increasing effective fluid velocity can enhance the force generation significantly at the onset of the stroke. However, the effectiveness of the wing-wake interaction mechanism strongly depends on the magnitude and the orientation of the wake vorticity during the stroke reversal [5], which are in turn obviously affected by the wing kinematics. It is clear that the contribution of wing-wake interaction will be diminished or completely absent in forward flight due to convection of the wake structures with respect to the wings; however, it offers a great potential for lift generation during hovering flight.

The experimental study performed by Dickinson et al. [14] verified the significance of wake capture mechanism in three-dimensional flapping wing aerodynamics. They observed two force peaks during the stroke reversal, one of which was attributed to the rotational circulation. Nonetheless, their tests revealed that the other peak appears just after the wing changes direction regardless of the phase of the wing rotation. In order to investigate the source of this force contribution, they halted the flapping motion at the end of the stroke arguing that if the source is the wake capture mechanism, the wing should still continue to generate force although it is at rest. As a result, they observed that the wing continues to generate forces and the force peak appears at the same position with the peak in the continuous flapping motion in the time line. They supported force measurements with flow visualizations via particle image velocimetry (PIV) at the instant of stroke reversal showing that the peak-induced velocities in the vicinity of the wing are sufficiently high to generate the measured forces. They also mentioned the influence of wing kinematics by showing that the advanced rotation in the stroke reversal phase generates higher force due to the wake capture effect that stems from a more energetic wake.

Birch and Dickinson [5] followed a different approach to investigate the wake capture mechanism. They performed simultaneous force and flow field measurements on a flapping wing of a dynamically scaled robot for a number of reciprocating cycles starting from rest. In order to isolate the aerodynamic influence of the wake capture mechanism, they subtracted forces and flow fields in the first stroke, from those of the fourth stroke. They argued that as the wake is just developing in the first stroke, it does not have a considerable effect on the generated forces so by using this methodology, the wake effects can be identified. Their approach revealed two major effects of the wake on the force generation: an initial augmentation that is associated with the wake capture mechanism; a following attenuation that is well correlated with the decreased effective angle of attack caused by downwash in the wake and well predicted by a quasi-steady model. The model, however, was not successful in calculating the wake capture related increase in the forces, although they took into account the instantaneous velocity field due to the unsteady nature of the wake capture phenomenon.

5.3.2.5 Added Mass

In the discussion of the LEV, rotational forces and clap-and-fling motion, it was shown that these mechanisms affect the circulation around the wing and in turn enhance force generation, therefore they are called circulatory forces. Different from these mechanisms, the wake capture mechanism is primarily based on the transfer of the fluid's momentum to the wing. The added mass effect also originates from a non-circulatory phenomenon and is also known as an acceleration reaction [12]. When a wing accelerates, it has to accelerate also the surrounding fluid. As a result, the wing experiences an inertial reaction force by the fluid that is being accelerated. It can be considered as an increase in the inertia of the flapping wing due to the mass of the accelerated fluid, i.e. added mass or virtual mass.

Acceleration effects can be very significant in flapping flight. In his early study, Osborne [44] attributed the large force coefficients in flapping insect flight to the acceleration forces. It was also claimed that the passive rotation, bending and twisting of the wing during stroke reversals are mainly due to inertial reaction forces [10,21]. Despite the importance of the acceleration reaction forces, it is not straightforward to estimate their relative contribution. This is because in most cases local accelerations of the fluid are not necessarily caused by the wing's acceleration. For instance time-dependent downwash induced by wake structures or fluid acceleration due to flow separation will cause the exertion of acceleration reaction forces on the wing.

The added mass of an accelerating thin wing is estimated as equal to the mass of fluid in an imaginary circular cylinder that is encompassing the wing with the chord length as its diameter [16]. Thus per unit span, the added mass is:

$$m' = \frac{1}{4}\rho \pi c^2 \tag{5.15}$$

The sectional lift contribution of the added mass term [52] for a specific case of an airfoil accelerating of the plunging and pitching motions normal to the free stream reads [16]:

$$L' = \frac{1}{4}\rho \pi c^2 \left[-\frac{d^2 y}{dt^2} - \frac{d^2 \alpha}{dt^2} c \left(\hat{x}_0 - \frac{1}{2} \right) + \frac{d\alpha}{dt} U_\infty \right] \tag{5.16}$$

where $d^2 y/dt^2$ is the plunging acceleration (negative downwards), $d^2\alpha/dt^2$ is the pitching acceleration and $d\alpha/dt$ is the pitching velocity. The first two terms on the right hand side of the equation stand for the added mass forces due to accelerations of the plunging and pitching motions, respectively. Note that the center of added mass is located at the mid-chord position which is reasonable as a pitching acceleration around the mid-chord axis will be zero as opposite contributions from the upper and lower half will cancel each other out. The last term is a quasi-circulatory lift which is a result of the virtual or apparent circulation generated due to the added mass rotating with the wing section [16]. This approach has been utilized in a number of studies in combination with the blade-element theory in order to estimate the added mass contribution from wing kinematics in three-dimensional flapping motions [31,46,50,59].

5.4 Conclusion

In this chapter, we tried to explain physical principles of fixed-wing and flapping-wing aerodynamics. Main force generation mechanisms of generic flapping flight are described briefly. The effectiveness of these mechanisms is discussed and their use in biological flapping flight is exemplified based on a number of studies from the literature. In the next chapter, we will focus on the aerodynamics of DelFly flapping-wing flight, which is at least equally complicated due to its inherent design aspects, such as non-isotropic flexible wing structure and four-wing flapping configuration with wings performing clap-and-fling motion.

References

1. I.H. Abbott, A.E. Von Doenhoff, *Theory of Wing Sections, Including a Summary of Airfoil Data*, Dover Books on Aeronautical Engineering Series (Dover Publications, New York, 1959)
2. O. Baskan, Experimental and numerical investigation of flow field around flapping airfoils making figure-of-eight in hover. Master's thesis, Middle East Technical University, 2009
3. J.J. Bertin, M.L. Smith, *Aerodynamics for Engineers* (Prentice Hall, New Jersey, 1998)
4. J.M. Birch, M.H. Dickinson, Spanwise flow and the attachment of the leading-edge vortex on insect wings. Nature **412**(6848), 729–733 (2001)
5. J.M. Birch, M.H. Dickinson, The influence of wing-wake interactions on the production of aerodynamic forces in flapping flight. J. Exp. Biol. **206**(13), 2257–2272 (2003)
6. R.J. Bomphrey, N.J. Lawson, G.K. Taylor, A.L.R. Thomas, Application of digital particle image velocimetry to insect aerodynamics: measurement of the leading-edge vortex and near wake of a Hawkmoth. Exp. Fluids **40**(4), 546–554 (2006)
7. R.J. Bomphrey, N.J. Lawson, N.J. Harding, G.K. Taylor, A.L.R. Thomas, The aerodynamics of Manduca sexta: digital particle image velocimetry analysis of the leading-edge vortex. J. Exp. Biol. **208**(Pt 6), 1079–1094 (2005)
8. R.J. Bomphrey, G.K. Taylor, A.L.R. Thomas, Smoke visualization of free-flying bumblebees indicates independent leading-edge vortices on each wing pair. Exp. Fluids **46**(5), 811–821 (2009)
9. J.F. Campbell, Augmentation of vortex lift by spanwise blowing. J. Aircr. **13**(9), 727–732 (1976)
10. T.L. Daniel, S.A. Combes, Flexible wings and fins: bending by inertial or fluid-dynamic forces? Integr. Comp. Biol. **42**(5), 1044–1049 (2002)
11. R. Demoll, Zuschriften an die Herausgeber. Der Flug der Insekten und der Vgel. Die Naturwiss. **27**, 480–482 (1919)
12. M.W. Denny, American Society of Zoologists Meeting. Air and water: the biology and physics of life's media (Princeton University Press, New Jersey, 1993)
13. M.H. Dickinson, K.G. Götz, Unsteady aerodynamic performance of model wings at low reynolds numbers. J. Exp. Biol. **174**, 45–64 (1993)
14. M.H. Dickinson, F.-O. Lehmann, S.P. Sane, Wing rotation and the aerodynamic basis of insect flight. Science **284**(5422), 1954–1960 (1999)
15. M.H. Dickinson, F.-O. Lehmann, K.G. Gotz, The active control of wing rotation by drosophila. J. Exp. Biol. **182**(1), 173–189 (1993)
16. C.P. Ellington, The aerodynamics of hovering insect flight. II: morphological parameters. Philos. Trans. R. Soc. Lond. Ser. B Biol. Sci. **1934–1990**(305), 17–40 (1984)
17. C.P. Ellington, The aerodynamics of hovering insect flight. III: kinematics. Philos. Trans. R. Soc. B Biol. Sci. **305**(1122), 41–78 (1984)
18. C.P. Ellington, The aerodynamics of hovering insect flight. IV: aeorodynamic mechanisms. Philos. Trans. R. Soc. Lond. Ser. B Biol. Sci. (1934-1990) **305**(1122), 79–113 (1984)
19. C.P. Ellington, The aerodynamics of hovering insect flight. V: a vortex theory. Philos. Trans. R. Soc. Lond. Ser. B Biol. Sci. (1934-1990) **305**(1122), 115–144 (1984)
20. C.P. Ellington, C. van den Berg, A.P. Willmott, A.L.R. Thomas, Leading-edge vortices in insect flight. Nature **384**(19/26), 626–630 (1996)

21. A.R. Ennos, The inertial cause of wing rotation in diptera. J. Exp. Biol. **140**(1), 161–169 (1988)
22. W.S. Farren, The reaction on a wing whose angle of incidence is changing rapidly. Rep. Mem. Aeronaut. Res. Comm. (Great Britain 1648, Aeronautical Research Committee, Great Britain, 1935)
23. Y. Fung, *An Introduction to the Theory of Aeroelasticity* (Dover Publications, New York, 1993)
24. I.E. Garrick, Propulsion of a flapping and oscillating airfoil. NACA: Report 567, National Advisory Committee for Aeronautics, Unites States (1937)
25. H. Glauert, The force and moment on an oscillating aerofoil. Rep. Mem. Aeronaut. Res. Comm. (Great Britain) 1561 (Aeronautical Research Committee, Great Britain, 1929)
26. K.O. Granlund, M.V. Ol, L.P. Bernal, Unsteady pitching flat plates. J. Fluid Mech. **733**, R5 (2013)
27. R.G. Grant, *Flight: The Complete History* (DK Publishing, London, 2007)
28. Ü. Gülçat, *Fundamentals of Modern Unsteady Aerodynamics* (Springer, Berlin, 2010)
29. R. Halfman, Experimental aerodynamic derivatives of a sinusoidally oscillating airfoil in two-dimensional flow. NACA: Technical Note 2465, National Advisory Committee for Aeronautics, Unites States (1951)
30. D.G. Hurley, The use of boundary-layer control to establish free stream-line flows. Adv. Aeronaut. Sci. **2**, 662–708 (1959)
31. T. Jardin, A. Farcy, L. David, Three-dimensional effects in hovering flapping flight. J. Fluid Mech. **702**, 102–125 (2012)
32. J.D. Anderson Jr, *Fundamentals of Aerodynamics* (McGraw-Hill, New York, 2001)
33. J. Katz, A. Plotkin, *Low Speed Aerodynamics: From Wing Theory to Pane Methods* (McGraw-Hill, New York, 1991)
34. V.M. Kramer, Die zunahme des maximalauftriebes von tragflugeln bei plotzlicher anstell-winkelvergrosserung (boeneffekt). Z. Flugtech Motorluftschiff **23**, 185–189 (1932)
35. P.K. Kundu, I.M. Cohen, *Fluid Mechanics*, 4th edn. (Elsevier, New York, 2008)
36. F.-O. Lehmann, The mechanisms of lift enhancement in insect flight. Naturwissenschaften **91**(3), 101–122 (2004)
37. F.-O. Lehmann, P. Simon, The aerodynamic benefit of wing-wing interaction depends on stroke trajectory in flapping insect wings. J. Exp. Biol. **210**(Pt 8), 1362–1377 (2007)
38. F.-O. Lehmann, S.P. Sane, M.H. Dickinson, The aerodynamic effects of wing-wing interaction in flapping insect wings. J. Exp. Biol. **208**(Pt 16), 3075–92 (2005)
39. D. Lentink, M.H. Dickinson, Rotational accelerations stabilize leading edge vortices on revolving fly wings. J. Exp. Biol. **212**(Pt 16), 2705–2719 (2009)
40. J.H. Marden, Maximum lift production during takeoff in flying animals. J. Exp. Biol. **130**, 235–238 (1987)
41. L.A. Miller, C.S. Peskin, Flexible clap and fling in tiny insect flight. J. Exp. Biol. **212**(19), 3076–3090 (2009)
42. F.T. Muijres, L.C. Johansson, R. Barfield, M. Wolf, G.R. Spedding, A. Hedenström, Leading-edge vortex improves lift in slow-flying bats. Science **319**(5867), 1250–1253 (2008)
43. M.M. Munk, Note on the Air Forces on a Wing Caused by Pitching. NACA: Technical note 191, National Advisory Committee for Aeronautics, Unites States, 1925
44. M.F.M. Osborne, Aerodynamics of flapping flight with application to insects. J. Exp. Biol. **28**(2), 221–245 (1951)
45. M. Percin, Flow around a plunging airfoil in a uniform flow. Master's thesis, Istanbul Technical University, 2009
46. M. Percin, B.W. van Oudheusden, Three-dimensional flow structures and unsteady forces on pitching and surging revolving flat plates. Exp. Fluids **56**(2), 1–19 (2015)
47. C. Edward, Predictions of vortex-lift characteristics by a leading-edge suction analogy. J. Aircr. **8**(4), 193–199 (1971)
48. E. Reid, *Airfoil Lift with Changing Angle of Attack* (Technical note, National Advisory Committee for Aeronautics, Unites States, Naca, 1927)

49. S.P. Sane, M.H. Dickinson, The control of flight force by a flapping wing: lift and drag production. J. Exp. Biol. **204**(Pt 15), 2607–2626 (2001)
50. S.P. Sane, M.H. Dickinson, The aerodynamic effects of wing rotation and a revised quasi-steady model of flapping flight. J. Exp. Biol. **205**(Pt 8), 1087–1096 (2002)
51. S.P. Sane, Review: the aerodynamics of insect flight. J. Exp. Biol. **206**(23), 4191–4208 (2003)
52. L.I. Sedov, *Two-dimensional Problems in Hydrodynamics and Aerodynamics* (Interscience Publishers, New York, 1965)
53. W. Shyy, Y. Lian, J. Tang, D. Viieru, H. Liu, *Aerodynamics of Low Reynolds Number Flyers* (Cambridge University Press, Cambridge Aerospace Series, Cambridge, 2007)
54. W. Shyy, H. Aono, S.K. Chimakurthi, P. Trizila, C.-K. Kang, C.E.S. Cesnik, H. Liu, Recent progress in flapping wing aerodynamics and aeroelasticity. Prog. Aerosp. Sci. **46**(7), 284–327 (2010)
55. A. Silverstein, U. Joyner, Experimental verification of the theory of oscillating airfoils. NACA: Report 673, National Advisory Committee for Aeronautics, Unites States, 1939
56. R.B. Srygley, A.L.R. Thomas, Unconventional lift-generating mechanisms in free-flying butterflies. Nature **420**(6916), 660–664 (2002)
57. G. Taylor, M.S. Triantafyllou, C. Tropea, *Animal Locomotion* (Springer e-Books, Springer, SpringerLink, Berlin, 2010)
58. T. Theodorsen, General theory of aerodynamic instability and the mechanism of flutter. NACA: Report 496, National Advisory Committee for Aeronautics, Unites States, 1935
59. Q.T. Truong, Q.V. Nguyen, V.T. Truong, H.C. Park, D.Y. Byun, N.S. Goo, A modified blade element theory for estimation of forces generated by a beetle-mimicking flapping wing system. Bioinspiration Biomimetics **6**(3), 036008 (2011)
60. Z.J. Wang, Dissecting insect flight. Annu. Rev. Fluid Mech. **37**(1), 183–210 (2005)
61. T. Weis-Fogh, Quick estimates of flight fitness in hovering animals, including novel mechanisms for lift production. J. Exp. Biol. **59**, 169–230 (1973)
62. J.Z. Wu, A.D. Vakili, J.M. Wu, Review of the physics of enhancing vortex lift by unsteady excitation. Prog. Aerosp. Sci. **28**(2), 73–131 (1991)

Abstract

In this chapter, an overview of the experimental studies focusing on the aerodynamics of the DelFly is given. Force generation mechanisms of the DelFly flapping-wing flight are addressed with particular consideration of the effects of different parameters such as wing geometry and flexibility. Furthermore, state-of-art flow visualization results are presented to give insight into the behavior of flow structures around and in the wake of the flapping wings and to assess their influence on the unsteady forces.

6.1 Introduction

Notwithstanding the increased interest in flapping-wing flight, there are still many open questions concerning the unsteady aerodynamics around flapping wings of flying animals and devices. The complexity of the system increases substantially as one progresses from a model of a two-dimensional rigid wing placed in a free-stream flow with a limited effect of viscosity toward a three-dimensional flapping flexible wing. The complexity is further enhanced by the two-way fluid-structure interaction due to wing flexibility: the wings do not only influence the surrounding air flow but are also deformed by the same stream. Clearly, the study of flapping-wing aerodynamics particularly with three-dimensional flexible wings is not straightforward and requires a systematic approach to properly identify effects and contributions of different elements.

In this regard, the identification and optimization of the DelFly force generation mechanisms have been performed in a number of separate studies that focused specifically on a single aspect of the complete mechanism by minimizing the effects of other parameters as much as possible. Accordingly, force and flow field measurements were carried out on the DelFly in different flight regimes for a number of experimental parameters and an overview of the results with some key conclu-

© Springer Science+Bussiness Media Dordrecht 2016

G.C.H.E. de Croon et al., *The DelFly*, DOI 10.1007/978-94-017-9208-0_6

sions are presented in this chapter. The organization of the chapter is as follows: first, force measurements in vacuum and in air conditions are compared in order to identify aerodynamic and inertial force components, which is complemented with the images from high-speed photography performed in the two conditions. Second, a brief discussion about the wing shape optimization study is included with the comparison of flow fields via particle image velocimetry (PIV) measurements around the initially designed and the optimized wings. Subsequently, effects of wing span length and the wing flexibility are investigated by means of force measurements in hover condition. Finally, the results of the three-dimensional wake reconstruction study by means of stereoscopic PIV (stereo-PIV) measurements at several streamwise locations in forward flight configuration are presented with a special focus on the effect of the reduced frequency on the flow structures and the force generation mechanisms.

6.2 Comparison of the Flapping Motion and Unsteady Forces in Air and Vacuum Conditions

Investigation of the aerodynamic performance of the DelFly flapping-wing motion in terms of force generation has been carried out by means of mounting a DelFly model on a force sensor and measuring temporal variation of forces at several selected flight operating conditions. However, the force sensor responds to aerodynamic forces and moments as well as to inertial effects and mechanical vibrations of the DelFly components, which makes it difficult to isolate the aerodynamic force components in the force spectrum and assess the real impact of the flapping-wing motion. In the sense of analysis of the mean forces, it is plausible to assume that the inertial force generated by the wing during upstroke is mostly canceled out by that during downstroke for a symmetrical wing motion kinematics, which results in a time-averaged value of zero. Nevertheless, proper analysis of the time-history of the aerodynamic forces requires more elaborate investigation to distinguish between the aerodynamic and inertial contributions to the resultant forces and moments. For this reason, in-vacuum force measurements were performed and the results are compared to those of the in-air measurements in this section.

The in-vacuum measurements were performed in the test section of the hypersonic wind tunnel at the Aerodynamics Laboratory of Delft University of Technology. The in-vacuum tests were carried out at a pressure level 1 mbar, whereas measurements in air were performed again in the test section with the same experimental configuration (except the pressure level in the test section) for the sake of a proper comparison of the two cases. Subsequently, in-air measurements were repeated out of the test section with the DelFly mounted on a balance system for a higher quality high-speed imaging of the DelFly flapping motion and to gain more reliable knowledge of the relation between the unsteady forces and the phase of the flapping motion.

(a)

(b)

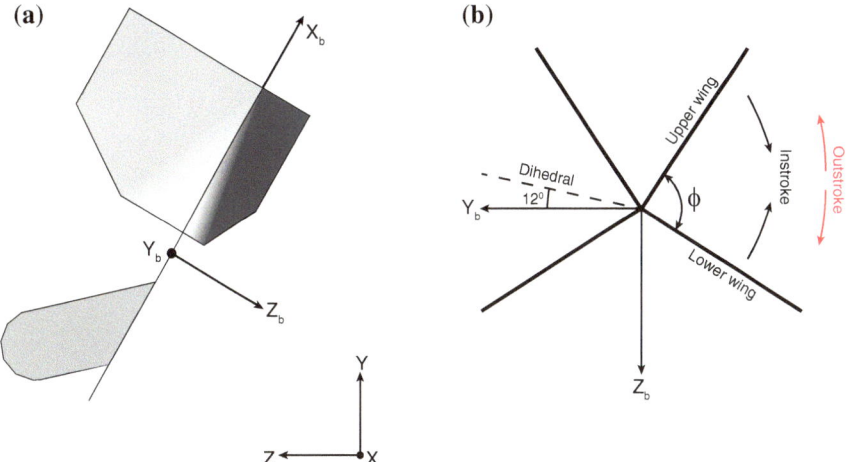

Fig. 6.1 **a** Inertial (X, Y, Z) and DelFly body (X_b, Y_b, Z_b) coordinate systems; **b** *front view* of the wing leading edges. Reprinted with permission

A full-scale DelFly II model without a tail in hovering flight configuration was used as the test model in the measurements. In this case, the DelFly is oriented vertically (x_b axis is parallel to y axis, see Fig. 6.1a), therefore the X-force that is oriented in the x_b direction of the body coordinate system is the focus of interest because it is the force component that balances the weight of the DelFly. Forces and moments were captured by the use of a six-component force sensor (ATI Nano-17 with a maximum sensing value of 32 N in x, y and 56.4 N in z direction with a resolution of 1/160 N and a measurement uncertainty of 1 % of the full-scale load with 95 % confidence level).

The synchronization and data acquisition system consists of two main subsystems: (1) an in-house developed microcontroller system that is mainly responsible for the regulation of the flapping frequency and the synchronization of the force measurements and high-speed imaging with the flapping motion; (2) an in-house programmed Field-Programmable Gate Array (FPGA) system of National Instruments that is responsible for the data acquisition. The microcontroller system counts the electrical commutations of the brushless DelFly motor for high-resolution rotational information and reads a Hall sensor with its magnet placed on the wing driving gear of the model for zero-referencing. Based on this information, the system then regulates the power supplied to the DelFly motor with a very slow integrator-only controller to keep the average flapping frequency constant on the long run. The FPGA system, on the other hand, acquires motor commutation information, the Hall sensor signal, voltage and current fed to the brushless motor as measured in the micro-controller system in addition to six components of forces and moments at a given data acquisition frequency. In the present experiments, the data were captured at a

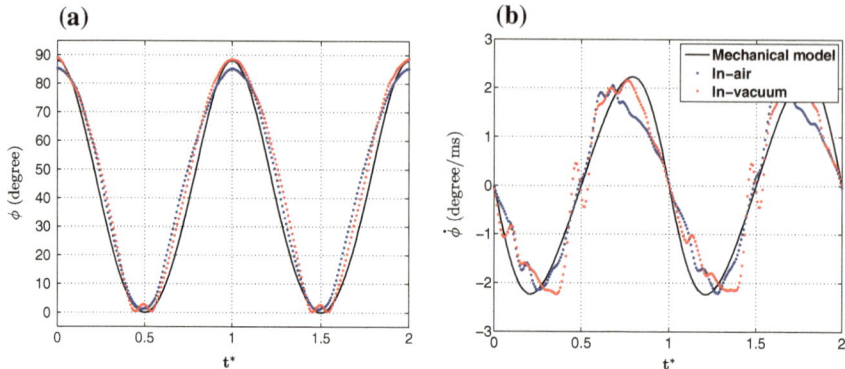

Fig. 6.2 Temporal variation of the stroke angle (ϕ) and angular velocity ($\dot{\phi}$) versus non-dimensional time (t^*) for two periods of the flapping motion at 8 Hz flapping frequency in air (*blue*) and in vacuum (*red*), complemented with the prescribed motion kinematics calculated based on the mechanical model of the driving system (*black*). Reprinted with permission

recording rate of 10 kHz. In the experiments, the DelFly flapping frequency was varied in the range of 8–12 Hz.

In order to obtain the motion kinematics and wing deformation characteristics, a Photron Fastcam SA 1.1 camera was placed to observe the wings of the DelFly model in front view. The flapping motion of the wings was recorded at a recording frequency of 1 kHz for a duration of 15 seconds to ensure that at least 10 flapping cycles captured even at the maximum flapping frequency. In the tests performed in the test section of the hypersonic wind tunnel, the background of the flapping wings was illuminated by a spotlight (FSP-575) to achieve a better contrast of the wing leading edges. Then, a dedicated MATLAB script was used to detect the leading edges and to calculate the stroke angle (ϕ), which is defined as the angle between the upper and the lower wings (Fig. 6.1b).

Figure 6.2a displays the temporal variation of the stroke angle plotted with respect to non-dimensional time ($t^* = t/T$, where t is time in seconds and T is the period of a flapping cycle) for two periods of the flapping motion at 8 Hz flapping frequency in air (blue) and in vacuum (red) conditions. Moreover, the variation of the stroke angle is calculated based on the mechanical model of the driving system, assuming a constant rotational speed of the driving axis and no deformations of the system (plotted in black). Note that the start of the instroke was selected to define the start of the flapping period.

The comparison of these three cases reveals significant differences in the kinematics. Initially the maximum stroke angle reaches approximately 85° in-air measurements, while it reaches 88° in-vacuum measurements, which agrees well with the design full stroke amplitude of the flapping system. This suggests that due to the aerodynamic loads exerted on the flexible wings, not only the flexible wing planform deforms but also the stiff leading edges bend and the driving hinge mechanism is constrained, which results in a decline of the maximum stroke amplitude of the flapping motion. Furthermore the motion profiles in air and vacuum conditions do not match with that of the mechanical design with a greater deviation in the flapping motion in air. The difference is also evident in the angular velocity profiles of the three cases, which is calculated by differentiating the phase angle. This most probably stems from the fact that the flapping frequency fluctuates at the sub-flapping-period level particularly during the in-air flapping motion because of the varying loads on the motor. Based on the motor commutation information, the standard deviation of these sub-period fluctuations was calculated as 13 and 5 % of the nominal flapping frequency for the in-air and in-vacuum measurements, respectively. In addition to these, the instroke lasts shorter than the outstroke ($t_{in}^* = 0.47T$) in air, which is also evident from the angular velocity profile (Fig. 6.2b) with a slightly higher instroke peak velocity (5 % relative difference). The absence of such a difference in the in-vacuum measurements suggests that it is most probably due to different flow structure characteristics caused by the wing-wing interaction taking place at the start of the outstroke (i.e. clap-and-peel). On the other hand, at the end of the instroke of the flapping motion in vacuum condition, there appears a small hump. This is due to the upper and lower wings hitting each other and bounce back for a stroke angle of approximately 2.5°. As the motion progresses, they approach until they gently touch each other this time and then start moving apart with the start of the outstroke. Apparently, in air, the aerodynamic force exerted on each wing counteracts the inertia of the wings so that the leading edges never get in full contact.

As a matter of fact, the most important aim of the force measurements in air and vacuum conditions is to identify the force components that are associated to the aerodynamics of the flapping-wing motion, while isolating the components that are originating from structural vibrations or wing inertia as they are also captured by the force sensor. Therefore, the influence of the flapping-wing aerodynamics on the force generation can be assessed and further optimization studies can be performed. For this purpose, power spectral densities of the X and Z force components in air and

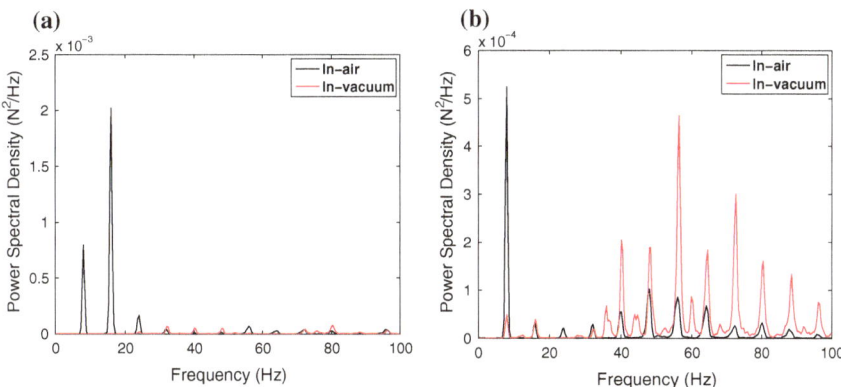

Fig. 6.3 Comparison of power spectral densities of **a** X and **b** Z forces in air and vacuum conditions for a flapping frequency of 8 Hz. Reprinted with permission

vacuum conditions are compared in Fig. 6.3 for the flapping frequency of 8 Hz. The spectrum of the X-force component in air contains two major peaks with frequencies corresponding to the first (fundamental) and second harmonics of the flapping motion in addition to a relatively less powerful third harmonic. None of these frequency components appear in the vacuum case which suggests that they can be attributed to the aerodynamic forces. For higher frequencies, the spectrum is relatively clean with very low-amplitude higher harmonics of the flapping frequency. On the other hand, the spectrum of the Z-force component is populated with high-harmonic peaks, which have a higher amplitude in vacuum and are significantly damped in air. Only the first harmonic, which is essentially the flapping frequency, emerges with a higher amplitude in air than that in vacuum conditions.

Flapping at a higher frequency (i.e. 12 Hz) results in slightly different power spectra (Fig. 6.4). In the spectrum of X-force, it is clear that the first (fundamental frequency) and the second harmonics of the flapping motion only appear for the case of in-air measurements, in accordance with the lower flapping frequency case. However, the third harmonic has the same amplitude both in air and vacuum conditions. This can be explained by the fact that for the flapping frequency case of 12 Hz, the frequency of the third harmonic probably corresponds a frequency of a structural mode of the DelFly body or one of the components. On the other hand, the forth and the sixth harmonics have higher amplitude in vacuum but are significantly damped in air. This suggests that these peaks likely correspond to wing structural mode excitations. In the spectrum of the Z force component, the first harmonic stands out with its amplitude higher in air than in vacuum conditions similar to lower flapping frequency case. On the other hand, the fourth, fifth and seventh harmonics present opposite characteristics with higher magnitude in vacuum conditions. This can be associated to the different structural and interaction characteristics of the wings driven by the flapping-wing inertia during the flapping motion.

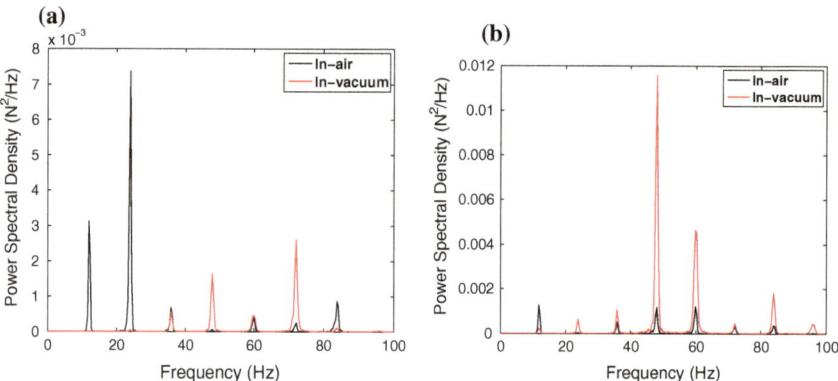

Fig. 6.4 Comparison of power spectral densities of **a** X and **b** Z forces in air and vacuum conditions for a flapping frequency of 12 Hz. Reprinted with permission

In view of force generation mechanisms of the flapping-wing motion, the second peak in the spectrum of the X-force (Fig. 6.4a) can be understood as the force generated by the translating motion of both wings that happens twice in one complete flapping cycle (during the instroke and outstroke phases) and associated delayed stall (leading edge vortex) mechanism. On the other hand, the first peak is related to the clap-and-peel motion, which can be regarded as a force enhancement mechanism, occurring only once during stroke reversal from instroke to outstroke in a flapping cycle. The third harmonic either has very low amplitude (in the low frequency case) or has similar amplitude in air and vacuum conditions (in the high frequency case). Apparently, in the latter case, the third harmonic of the flapping motion excites one of the body structural modes of the DelFly. Clearly, in the flapping frequency range of force measurements, the first two harmonics correspond to aerodynamic forces. Therefore, the raw X-force data was filtered by a Chebyshev II filter with -80 dB attenuation of the stop-band in MATLAB. The cut-off frequency was selected to keep only the first two harmonics while eliminating all higher modes, which are presumably associated to the wing and body structural modes of the DelFly.

Analysis of the Z force in hover condition is more difficult as in principle, the forces generated by the wings moving in opposite phase mostly (not completely due to dihedral angle of the wings) cancel each other resulting in a relatively small net Z-force production. Comparison of the Z-force in-air and in-vacuum conditions by means of power spectral densities reveals that the first harmonic is the only component that has a higher amplitude in air conditions pointing to aerodynamics related contribution for the specific case of hovering flight.

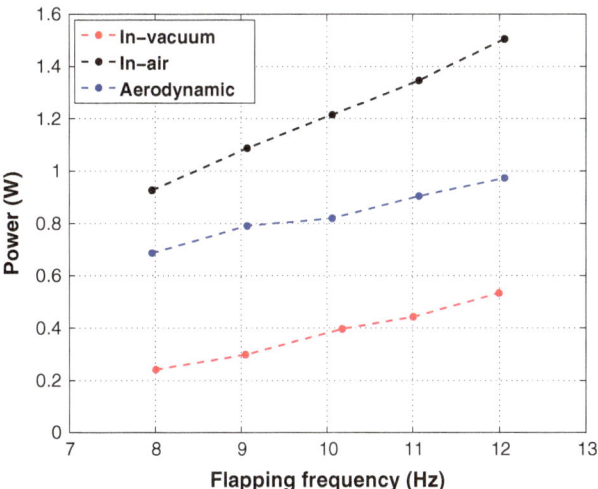

Fig. 6.5 Power consumption versus flapping frequency for the in-vacuum and in-air measurements complemented with the aerodynamic power computed as the total power consumption in air minus that in vacuum. Reprinted with permission

The power consumption that is required to sustain the DelFly flapping motion is calculated by multiplying time-resolved measured values of voltage and current fed to the brushless motor. The time-resolved values are then averaged over a number of flapping periods for the in-vacuum and in-air measurements for the flapping frequencies considered in the measurements (Fig. 6.5). Furthermore, the power consumption associated with the aerodynamic effects is computed as the total power consumption in air minus the power consumption in vacuum.

It is clear that the greater portion of the total power consumption (64–74 %) is due to aerodynamic effects. Overall power consumption in air follows a linearly increasing trend with respect to flapping frequency. Considering that the power required to drive the motor, gears and hinges is varying in the range between 0.18 and 0.2 W (based on complementary measurements on the DelFly flapping without the wings), the power required to overcome the elastic-inertial forces reaches about 20 % of the total power consumption in air for the case of the highest flapping frequency.

The comparison of spectrum of the forces in air and vacuum conditions have allowed to distinguish the aerodynamic and inertial components in the measured forces to some extent, as well as giving clues about the contributions of aerodynamic force generation mechanisms. However, clear identification of the aerodynamic mechanisms requires analysis of the data in a time-resolved manner. For this reason, temporal variations of the X-force and power consumption are plotted for two periods of the flapping motion for the case of flapping at 12 Hz in hovering configuration of the DelFly in air (Fig. 6.6-top). In order to assess the relation between the resultant forces and the flapping motion, images of the flapping wing-pair at six selected phases of the flapping cycle that correspond to local extremes of the X-force and power consumption are also provided in Fig. 6.6 (bottom).

For this specific case, the flapping wings generate a flapping-cycle-averaged X-force of 0.177 N, which is sufficient to support the weight of a DelFly with a typical mass of 17 g in hovering configuration. At the start of the instroke, the X-force is at the minimum of the complete flapping period. At the instant A, which corresponds to stroke reversal, the upper and lower wing leading edges are more or less stationary and the wing surfaces have a small amount of deformation. As the upper and lower wings start moving in the instroke direction, the force increases gradually with increasing wing speed and deformation. The maximum force during the instroke is attained at about $t^* = 0.3$, when the wing has the maximum stroke velocity. Furthermore, the wing surfaces are deflected considerably resulting in an increased frontal projected wing area with its normal in the X_b direction. Clearly, in terms of X-force production, such an area is required for the pressure fields to exert on. For instance, the low pressure region formed by the LEV requires a vertical wing surface area to exert on in order to produce a suction force. The power consumption during the instroke peaks slightly after the maximum of the X-force (at $t^* = 0.35$, instant C). In this period of the instroke, the stroke velocity is decreasing which causes also a decrease in the magnitude of the resultant forces. In agreement with this, the power consumption also decreases. Subsequently, the second minimum of the X-force is reached (at $t^* = 0.49$, instant D) at the end of the instroke. However, the force value is significantly higher at this stage than at the end of outstroke. In

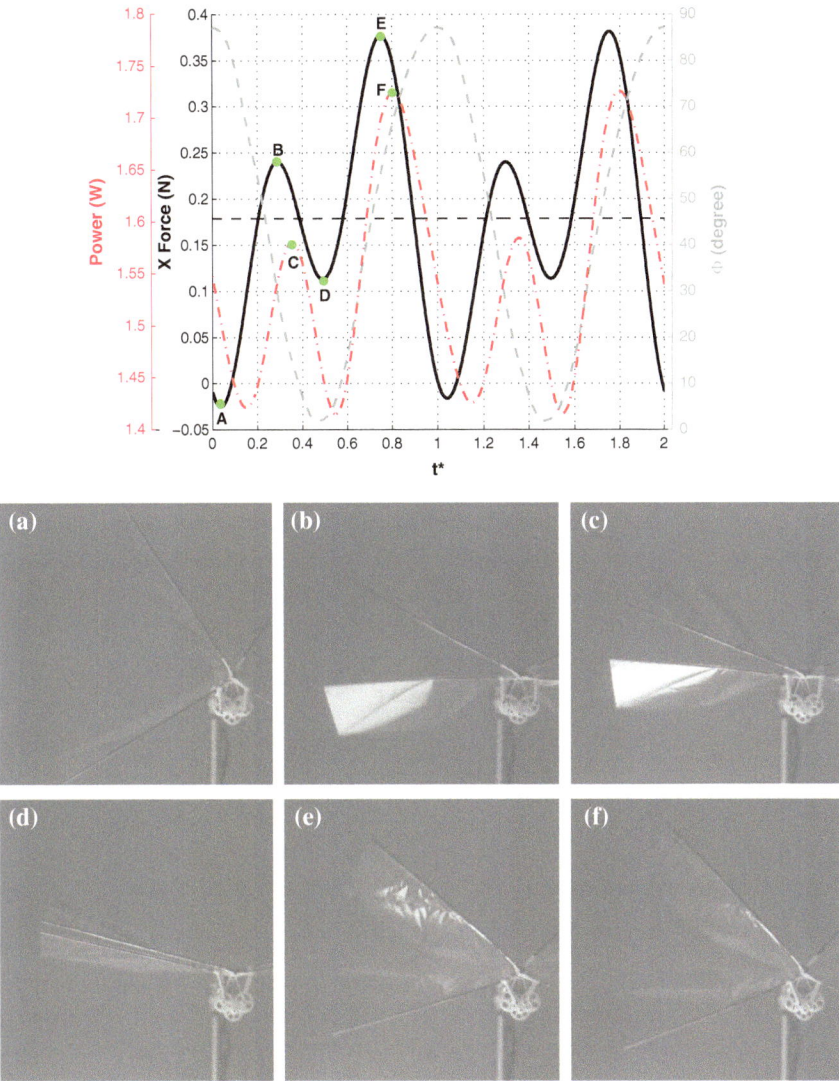

Fig. 6.6 Temporal evolution of the X-force (*black*) and power consumption (*red*) for two periods of the flapping motion for the flapping frequency of 12 Hz in hover configuration of the DelFly in air conditions complemented with the variation of the stroke angle (Φ, *dashed blue*) (*top*); six instants of the flapping cycle (*bottom*) which correspond to local maximums and minimums of the X-force and power consumption variations as indicated by letters A-F in the plot above. Reprinted with permission

this case, the wings clap and presumably the momentum jet generated due to this wing-wing interaction contributes to the X-force production. Also note that the total vertical projected area of the wings is larger at this time than at the end of the outstroke, which provides a relatively larger area for the suction forces. As the peeling occurs at the start of the outstroke, the X-force starts increasing again reaching finally the maximum of the complete flapping cycle at $t^* = 0.75$ (instant E). Note that at this very stage, the maximum stroke velocity has already been passed and it is in the decreasing phase. The high speed image of this instant shows that although the wing leading edges are apart at a stroke angle of approximately 56°, the trailing edges are still in contact until the taper point of the wings. This configuration promotes the formation of a low pressure region between the wings and the inrush of the fluid to this gap. In addition to the elevation of the X-force, this configuration also results in the increase of the power consumption because the low pressure region between the wings increases power requirement to peel the wings apart. The power consumption peak is off-phase with respect to the maximum X-force generation instant during the outstroke ($t^* = 0.8$, instant F). The phase lags between the maximums of the force generation and power consumption are likely to originate from the structural deformation characteristics of the flapping wings.

Force measurements in air and vacuum conditions together with high-speed imaging of the DelFly flapping motion allowed us to assess the aerodynamic and inertial contributions in the resultant measured forces and to associate the force generation with the different phases of the flapping motion. In the next section, a wing geometry optimization study is presented that aims to maximize the aerodynamic force to power consumption ratio during flapping motion in hovering flight.

6.3 Optimization of the Wing Geometry

The flapping motion of the DelFly wings consists of three main motion kinematics: first, the sweeping motion that is driven by the gear and hinge mechanism actively and can be varied in frequency but constrained in amplitude; second, the small-amplitude heaving motion that occurs due to bending of the leading edge; third, the pitching motion of the wings that occurs passively mostly during stroke reversals as a result of interaction between the aerodynamic, inertial and elastic forces acting the wings. These force components also play an important role on the determination of the angle of attack during the instroke and outstroke phases of the flapping motion. Obviously, the fluid-structure interaction is an important aspect of the flapping flight of the DelFly in terms of force generation and power consumption. One way to tune this interaction in order to increase the efficiency is to change the wing geometry and stiffness of the wings, which have either direct or indirect influence on the aerodynamic, inertial and elastic force components.

However, the space of all possible wing geometries is vast. As a consequence, any empirical study of the wing geometry will have to limit the considered parameters. In this respect, the most stringent limiting factors in the design of an MAV can be considered as the maximum length scale and the maximum overall weight. For this reason, in the presented wing geometry study, wing span length and the wing area were kept constant. Instead, the study focused on the locations and orientations of the stiffeners on the wing, as this was expected to have a considerable effect on the resultant aerodynamic force generation and power consumption by changing the stiffness of the wings. As a measure of the aerodynamic performance, the ratio of X-force to power consumption (X/P) was used. The raw force data was filtered with a low-pass filter that has a cut-off frequency of twice the flapping frequency to account for only aerodynamic force fluctuations. The experimental results reported in this section are based on the master thesis studies performed by Bruggeman [5] and Groen [11].

6.3.1 Influence of the Stiffener Position

In these tests, the orientation of the stiffeners was systematically approached. First, the two stiffeners were positioned parallel to each other under five different angles with the horizontal axis (defined positive in the clockwise direction): $0°$, $31°$, $63°$, $77°$ and $90°$. Note that the tested wings are named based on the angles of stiffeners. In case the stiffeners are parallel to each other, the wing is coded with a single number of the angle of both stiffeners. Otherwise, the wing name consists of two numbers: the first two digits represent the angle of the inwards stiffener (the one close to the wing root) and the last two digits correspond to the angle of the outwards stiffener (the one close to the wing tip), respectively. The placements of the stiffeners are depicted in Fig. 6.7. Each configuration was tested for 20 s at different flapping frequencies. The presented experimental values of force and power consumption are the average values over that time interval.

The results of these tests are presented in Fig. 6.8. Obviously higher X-force-to-power ratio means more force generation for a unit power consumption that indicates a superior performance of the flapping-wing motion. It is clear that *wing31* and *wing63* have relatively high X/P for the complete range of test flapping frequencies. In general, the maximum force is generated by *wing31* at an expense of the maximum

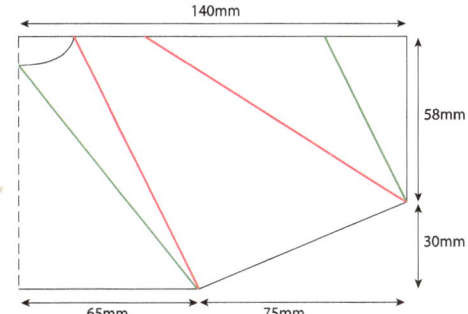

Fig. 6.7 Schematic representation of the half wing with stiffener orientations of *wing0* (*cyan*), *wing31* (*magenta*), *wing63* (*blue*), *wing77* (*green*), *wing90* (*red*) and the standard wing (*wing5424*, *dotted*), which were tested in the first measurement campaign (adapted from [5])

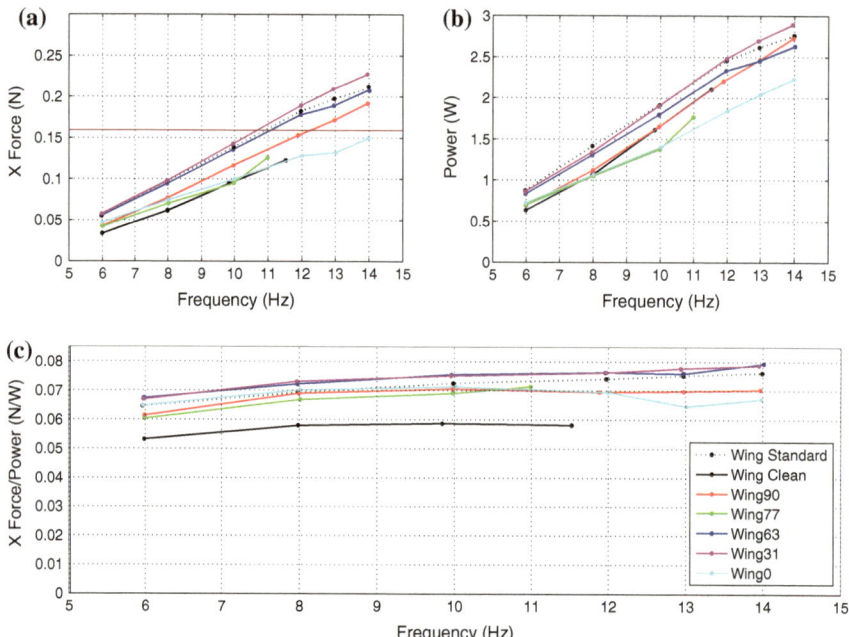

Fig. 6.8 Flap-averaged **a** X-force (the *brown line* represents the DelFly II weight of 0.16 N), **b** power consumption and **c** X-force-to-power ratio versus flapping frequency for the cases of different stiffener orientations for the first measurement campaign as depicted in Fig. 6.7 (adapted from [5])

power consumption. The wing without stiffeners (Wing Clean) has poor characteristics resulting in the minimum X/P ratio for a given flapping frequency. On the other hand, *wing0* outperforms the clean wing although both of them produce equivalent force for the flapping frequencies above 9 Hz. From a structural point of view, the stiffeners in the case of *wing0* do not stiffen the wing in the chordwise direction but contribute to the spanwise stiffness, which is mainly provided by the relatively thick and rigid leading edge. The presence of the stiffeners, however, changes the value and the distribution of the mass of the wings, which in turn influences the inertial forces. As a consequence, *wing0* consumes relatively less power with respect to other configurations, which can be attributed to the enhanced chordwise wing deformation and accordingly decreased drag because of smaller wing area facing the direction of the wing sweeping motion.

As a result of the initial tests, *wing31* and *wing63* proved to be the most efficient configurations. In order to improve the performance even more, combinations of these wing are built, viz., *wing6331* and *wing5163* (Fig. 6.9). Note that for the latter, the angle of the inwards stiffener was increased to 51° because otherwise the stiffener penetrates through the foil at the root side. This was also observed in the experiments with *wing31* so that an adapted version of *wing31* was built by changing the angle of the inner stiffener (*wing5131*).

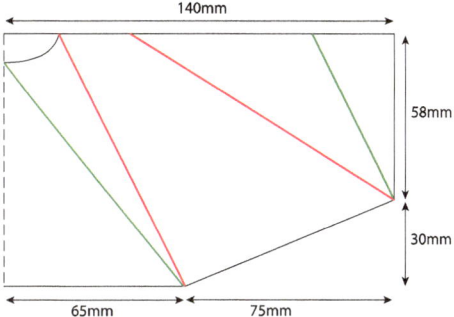

Fig. 6.9 Schematic representation of the half wing with stiffener orientations of *wing6331* (*red*) and *wing5163* (*green*), which were tested in the second measurement campaign (adapted from [5])

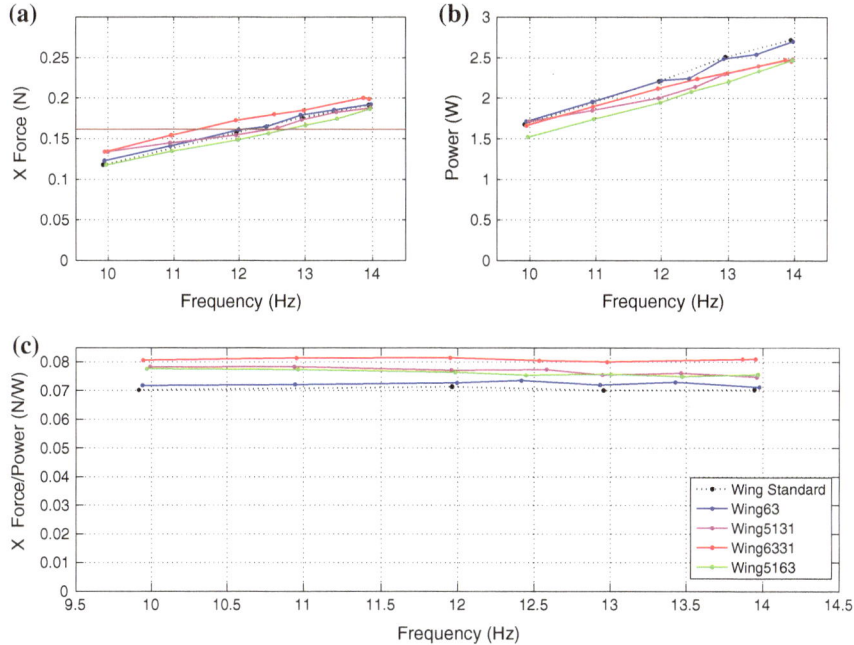

Fig. 6.10 Flap-averaged **a** X-force (the *brown line* represents the DelFly II weight of 0.16 N), **b** power consumption and **c** X-force-to-power ratio versus flapping frequency for the cases of different stiffener orientations for the second measurement campaign as depicted in Fig. 6.9 (adapted from [5])

One can notice that the standard wing and *wing63* have slightly different performance in the second test campaign than in the first campaign (Fig. 6.10) due to the wear of the flapping mechanism. However, the relative difference between the performance measures of different wings is the same for each test campaign as the measurements were performed successively. The analysis of the second set of measurements reveals that *wing6331* outperforms the other configurations in all measures. Moreover, it was noticed that wings with converging stiffeners toward the leading edge perform better than the wing with diverging stiffeners. With this in mind, number of wings with converging stiffeners at various angles were built to be tested in the third test campaign.

Starting from the winner of the second test campaign, i.e., *wing6331*, and by systematically changing the angles of the stiffeners, eight different configurations all with converging stiffeners toward the leading edge were built for the third measurement campaign (see Fig. 6.11).

Among the tested configurations, *wing6331* (configuration 1) generates the highest X-force but also has the highest power consumption resulting in a relatively low X-force-to-power ratio (Fig. 6.12). *Wing7731* (configuration 4) has the best performance in terms of X/P, which is followed by *wing9039* (configuration 7). Apart from these, *wing11663* (configuration 8) in which the stiffeners intersect closer to the wing tip performs poorly by generating the lowest force and having the minimum X/P for the complete range of test flapping frequencies.

As a final step to improve the aerodynamic performance of the flapping wings, the best two configurations of the third test campaign were picked (*wing7731* and *wing9039*) to generate another wing configuration, namely *wing8435* in which the convergence point of the stiffeners toward the leading edge is in the middle of those of the *wing9039* and *wing7731*. Force and power consumption measurements revealed superior performance of this last configuration both in terms of force generation

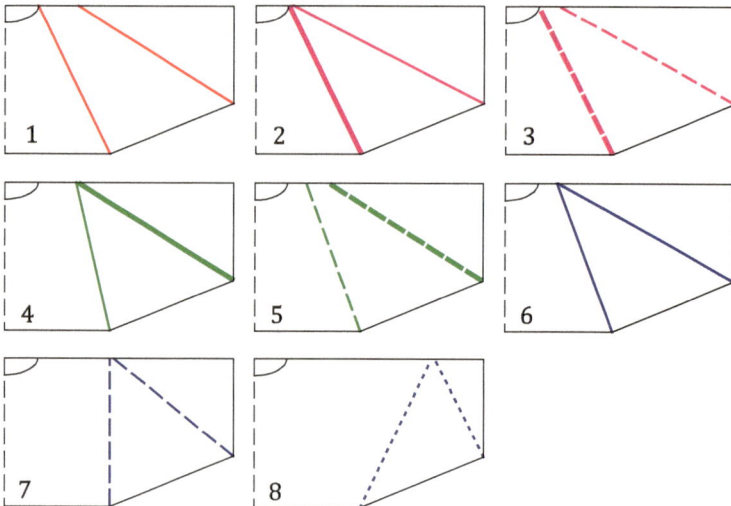

Fig. 6.11 Schematic representation of half wing with stiffener orientations of **a** *wing6331* (*red*), **b** *wing6326* (*solid magenta*), **c** *wing6328* (*dashed magenta*), **d** *wing7731* (*solid green*), **e** *wing6931* (*dashed green*), **f** *wing6928* (*solid blue*), **g** *wing9039* (*dashed blue*), **h** *wing11663* (*dotted blue*), which were tested in the third measurement campaign (adapted from [5])

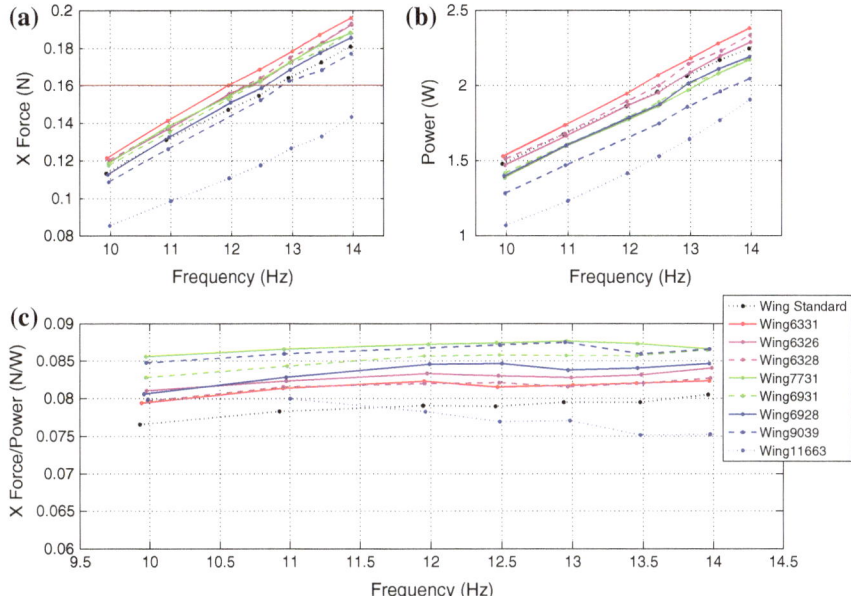

Fig. 6.12 Flap-averaged **a** X-force (the *brown line* represents the DelFly II weight of 0.16 N), **b** power consumption and **c** X-force-to-power ratio versus flapping frequency for the cases of different stiffener orientations for the third measurement campaign as depicted in Fig. 6.11 (adapted from [5])

and force-to-power ratio. As a result, *wing8435* (see Fig. 6.14b) was selected as the DelFly wing configuration and has been used in all applications of the DelFly II since 2010. The standard wing configuration, which was inherited from the older brother DelFly I, was abandoned due to its relatively poor aerodynamic performance. Comparative analysis of the new and the old wing configurations in terms of force generation, power consumption, wing deformation and flow field structures is given subsequently. Yet, before moving into this performance comparison, the effect of another property of the wing stiffeners, i.e., the stiffener diameter, is discussed in the next section.

6.3.2   Influence of Stiffener Diameter

The influence of stiffener diameter (and therefore also the influence of its stiffness) is investigated by changing it from 0.28 mm, which is used as the standard size in all DelFly II wings, up to a maximum of 1.0 mm, see Table 6.1.

The difference in inner and outer diameter of the stiffener will result in a different deformation during flapping. This can be translated to a difference in force generation and power consumption. From Fig. 6.13, it can be observed that the X-force and power consumption increase with the increase of cross-section area of the stiffener.

Table 6.1 Stiffener properties

Diameter (mm)	Wing mass (g)
0.28	1.1
0.5	1.3
0.7 (outer) −0.3 (inner)	1.4
1.0 (outer) −0.3 (inner)	1.7
1.0	2.2

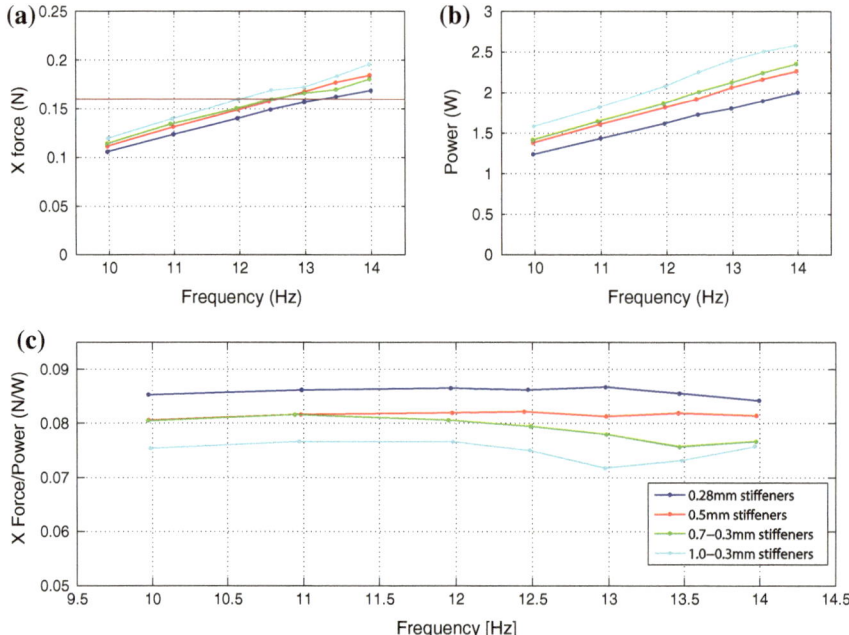

Fig. 6.13 Flap-averaged **a** X-force (the *brown line* represents the DelFly II weight of 0.16 N), **b** power consumption and **c** X-force-to-power ratio versus flapping frequency for the cases of different stiffener thicknesses (adapted from [5])

The X-force-to-power ratio, however, decreases with increasing cross-sectional area. In this respect, the wing with 0.28 mm stiffeners has the best characteristics. The wing with 1.0 mm diameter stiffeners is not plotted in Fig. 6.13 because the mechanism failed at wing flapping frequencies of 10 Hz and higher. This wing is relatively heavy in comparison to the other wings and therefore the motor controller was unable to control the wing flapping frequency due to high inertial forces. As a result, the wing stiffener thickness of 0.28 was kept as the standard size to be used in the construction of the DelFly II wings.

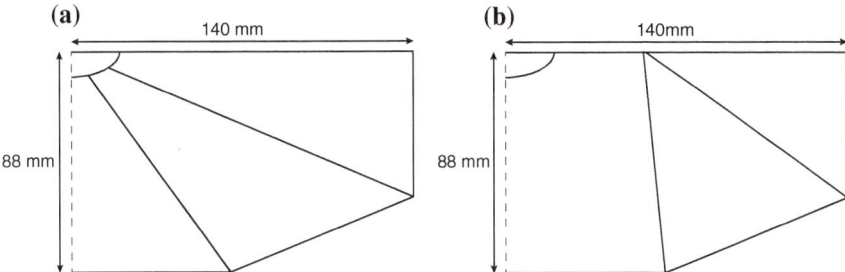

Fig. 6.14 Stiffener location and orientation of **a** the original wing and **b** the improved wing (adapted from [5])

6.3.3 Comparison of the Original and the Improved Wing

The wing geometry optimization study in terms of stiffener orientation and thickness led to the improved wing, which has the same area and the shape with the original wing but differs in the placement of the stiffeners (Fig. 6.14). In this section, first we will compare the performance of the initial and the improved wing by means of force and power consumption measurements. Then, in-flight shape of the two wings are compared via visualization of the wing cross-sections at a spanwise location. Finally, results of stereo-PIV measurements are presented, which show the formation and the behavior of flow structures in the vicinity of the flapping wings.

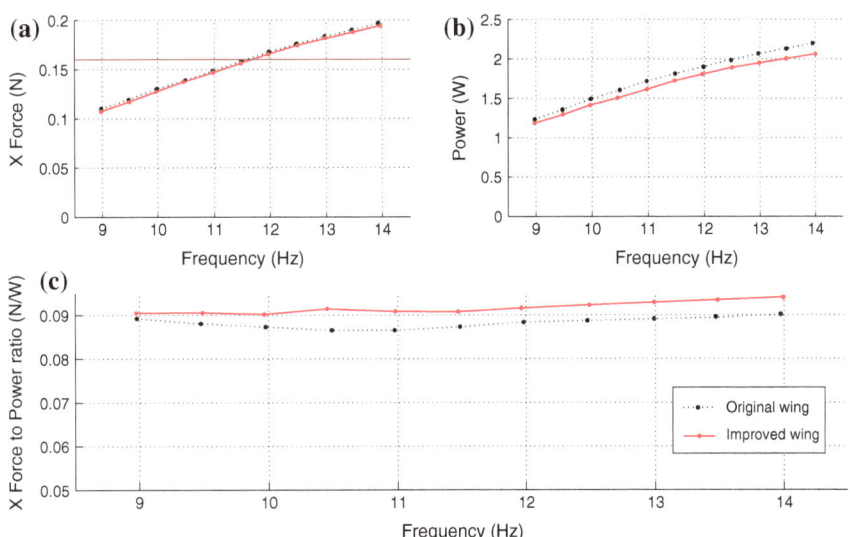

Fig. 6.15 Flap-averaged **a** X-force (the *brown line* represents the DelFly II weight of 0.16 N), **b** power consumption and **c** X-force-to-power ratio versus flapping frequency for the cases of the original and the improved wings (adapted from [5])

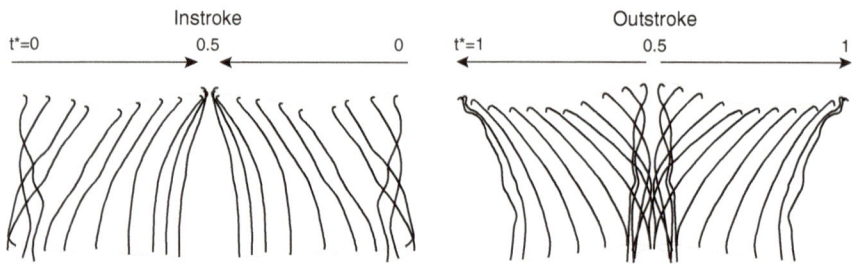

Fig. 6.16 Cross-sections of the original wing during a flap cycle at a flapping frequency of 11 Hz at a spanwise location of $0.71R$ (Loaded with an average X-force of 0.15 N) (adapted from [11])

In terms of X-force generation, the two wings perform almost equally with the original wing producing slightly higher force (Fig. 6.15). However, the improved wing clearly consumes less power than the original wing for a given flapping frequency. Consequently, the improved wing exhibits a more favorable X/P in the test flapping frequency range with an average increase of 5 %.

Changing stiffener location and orientation from the original to the improved wing naturally brings forth different wing deformation characteristics. As discussed previously, the in-flight wing shape is determined by the aerodynamic, elastic and inertial forces. The aerodynamic forces are in turn affected by the wing shape, leading to a complex fluid-structure interaction. Therefore, assessment of the in-flight wing deformations is crucial to explain aerodynamic effects.

As will be discussed subsequently, the air flow around the wings is investigated by means of stereo-PIV measurements, which allows the in-flight wing shape to be extracted from the images taken with the PIV cameras. Using the PIV setup at a low laser intensity (to minimize the reflections from the wing surface) without seeding particles, the cross-section of the wing at the spanwise position of $0.71R$ was illuminated. In Fig. 6.16, the shape of the original DelFly wing is shown at several instants during a cycle of the flapping motion at 11 Hz as a function of non-dimensional time (t^*). The cross-sections show the foil folded over the D-shaped leading edge carbon rod. The orientation of the carbon rod provides rigidity in the sweeping direction but allows the leading edge to bend up and down in the chordwise direction more easily. This enables the wing to perform a small-amplitude heaving motion during the flapping motion, as shown by the leading edge path in Fig. 6.16. Another aspect is the occurrence of the clap-and-peel motion between the two wing pairs. During the outstroke, particularly from $t^* = 0.5$ to $t^* = 0.8$, while the wing surfaces peel apart in accordance with the motion of the leading edges, the wing foil still claps at the trailing edge. Since the leading edges are the part of the wing being driven by the flapping mechanism, the wing motion can be regarded as the prescribed displacement of the leading edge with the rest of the wing being dragged behind, analogous to a waving flag. This illustrates the importance of the wing flexibility in terms of resultant wing shapes and in turn aerodynamic force generation.

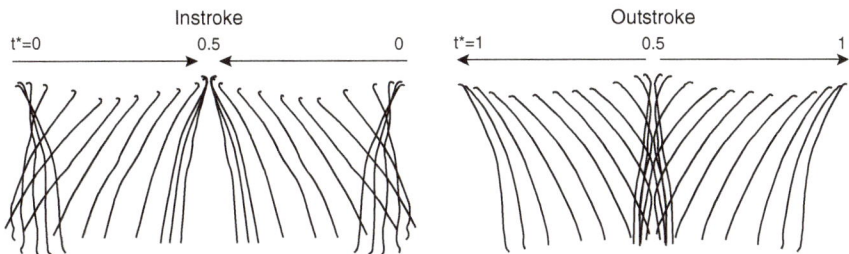

Fig. 6.17 Cross-sections of the improved wing during a flap cycle at a flapping frequency of 11 Hz at a spanwise location of $0.71R$ (Loaded with an average X-force of 0.14 N) (adapted from [11])

The comparison of the in-flight wing deformations (Fig. 6.17) reveals that the original wing behaves more flexible than the improved wing particularly during the stroke reversals. This is mostly because of the placement of the stiffeners being more outwards in the improved wing, which provides more rigidity at the locations near the wing tips. The more flexible behavior of the original wing during the stroke reversals could also originate from a difference in foil tension due to differences in the mounting of the wings to the DelFly body. Foil tension is difficult to control accurately because of sub-millimeter tolerances and the strong aging effect. On the other hand, the shape of both wings during the translation phases after the stroke reversals (both instroke and outstroke) is comparable.

The flow field around the flapping wings of the DelFly in hover flight condition was studied via stereo-PIV technique, which provides all three components of the velocity vector measured in a plane illuminated by a laser. The in-plane velocity components are used to investigate the vortex dynamics in the cross-sectional plane normal to the wing leading edge. The out-of-plane component represents the spanwise flow velocity in this configuration. The measurements were performed in a phase-locked manner with the triggering of the PIV system and its synchronization with the flapping motion being carried out by the microcontroller system. The velocity fields were acquired in total 34 phases of a flapping cycle for three different flapping frequencies (viz., 9, 11 and 13 Hz) at five different spanwise locations of the imaging plane, ranging from $0.42R$ to $1.0R$. For a detailed description of the experimental setup, the reader is referred to [11, 12].

Investigation of the vortex development requires their identification and quantification. The latter can be performed by use of calculation of vorticity (curl of the velocity vector field). Nevertheless, vortex core detection cannot be fulfilled successfully by means of vorticity at all time as vorticity also recognizes shearing motion within the flow. For this reason, the swirling strength [1] is utilized to determine the vortex cores and vortex strength. The swirling strength of a local swirling motion is quantified by γ_{ci}, which is the positive imaginary part of the eigenvalue of the local velocity gradient tensor.

Although the flow field measurements were performed for a number of different wings at several spanwise positions and flapping frequencies [11], only the results for the improved and the original wings at a spanwise position of $0.71R$ for the flapping

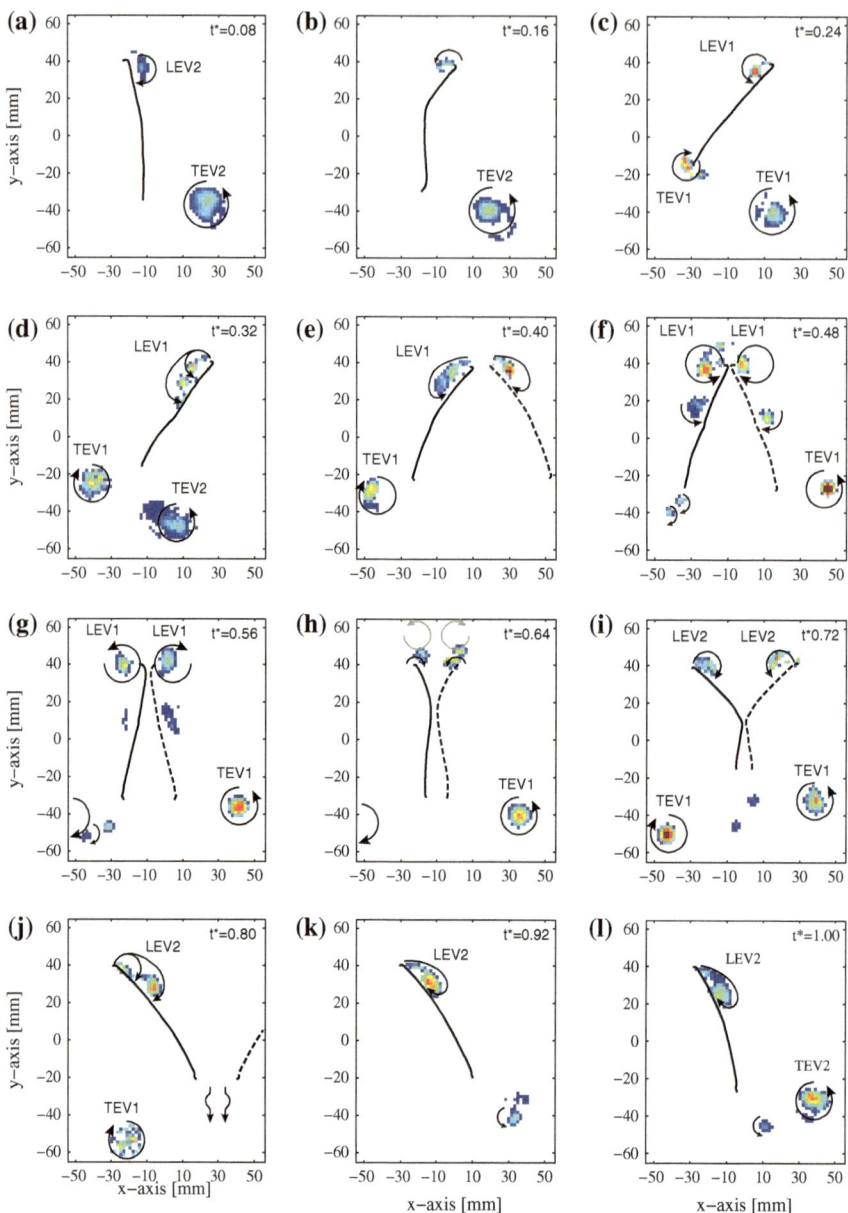

Fig. 6.18 Swirling strength at the spanwise location of $0.71R$ at various instants during a flap cycle for the improved wing flapping at 13 Hz complemented indications for the LEVs and TEVs generated during instroke (1) and outstroke (2) (adapted from [11])

frequencies of 11 and 13 Hz are discussed in this section. Figure 6.18 illustrates the development of leading edge vortices (LEVs) and trailing edge vortices (TEVs) during a flapping cycle. At the beginning of the instroke (a), the LEV and TEV from the previous outstroke are present in the flow field. As the wing proceeds moving inwards, the LEV of the outstroke disappears and a new LEV starts to form (b). At approximately half-way instroke (c), the LEV and the TEV are present in the flow field which are rather attached to the wing surface. With the progress of the motion, these two vortices grow larger both in size and strength with the TEV located further downstream from the trailing edge and the LEV being spread over the wing surface but still staying attached to the curved body (d). This pattern is preserved until the leading edges touch each other (f), when the LEV presents a scattered structure with a small scale shed LEV present approximately at the half-chord position and a relatively large vortex staying still around the leading edge. During the stroke reversal, the LEV of the instroke circumnavigates around the leading edge and interact the newly formed LEV of the outstroke (g–h). As the fling occurs, the LEV of the instroke grows larger (c) and subsequently expands over the chord-line toward the trailing edge (i–j). At the end of the outstroke (l) when the wing decelerates for rotation, the LEV decreases in strength.

In terms of the development of the LEV, both instroke and outstroke have similar patterns. The LEV during the outstroke appears stronger and closer to the wing surface, which can be attributed to the formation of downward velocity region induced by the peeling motion of the wings. These features of the LEV could account for the elevated forces of the outstroke phase. Moreover, during the stroke reversal from instroke to outstroke, the LEVs from the instroke remain around the leading edge and interact with those of the outstroke, whereas the LEV presumably advects toward the trailing edge and diffuses during the reversal from the outstroke to the instroke. During both instroke and outstroke also a TEV (starting vortex) is generated. However during the outstroke, the formation of the TEV is postponed due to clap-and-peel wing interaction phase. As the trailing edges separate from each other after the peeling phase of the motion, the TEV appears to start from a complex fluid structure (j).

Examination of the flow fields along the wing span reveals that the LEV develops conically with an increasing size toward the wing tip. The LEV appears first at the outwards stations, where the translational velocity is relatively high and at a later phase of the flapping period at the inwards positions. The LEV tube does not extend all the way to the wing tip. The LEV is still clearly visible at $0.86R$ (see Fig. 6.19), after which it disappears presumably bending toward the trailing edge connecting to the tip vortex. Inspection of the spanwise velocity component shows a spanwise flow toward the wing tip inside the vortex tube. The magnitude of this velocity component reaches approximately the translational velocity of the wing at the given spanwise position.

Comparison of the flow fields of the original and the improved wings nearly halfway during the outstroke (Fig. 6.19) reveals prominently different LEV structures: a larger LEV, which is detached from the surface, is present in the case of the original wing; while the improved wing features an LEV that is spread and attached on the wing surface. Such a difference can be explained by the different downward

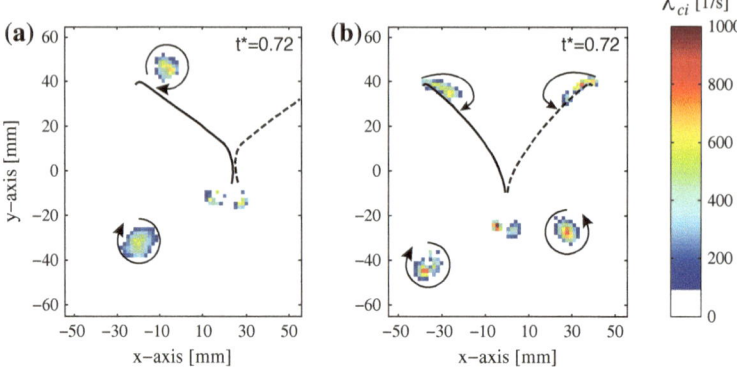

Fig. 6.19 Swirling strength at the spanwise location of $0.86R$ of **a** the original and **b** the improved wings at $t^* = 0.72$ for the flapping frequency of 11 Hz (adapted from [11])

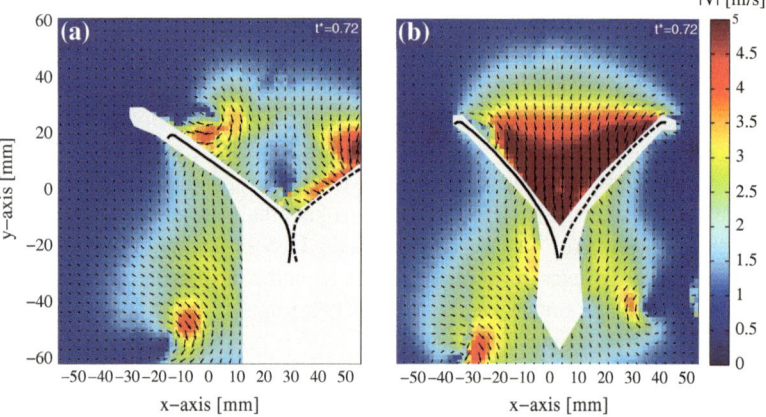

Fig. 6.20 Velocity vector field with the contours of in-plane velocity magnitude at the spanwise location of $0.86R$ of **a** the original and **b** the improved wings at $t^* = 0.72$ for the flapping frequency of 11 Hz (adapted from [11])

flow patterns in between the peeling wings for the two cases, as evident in the contour plots of in-plane velocity magnitude in Fig. 6.20. Apparently, rotation of the more rigid improved wing increases the peeling effect and thus promotes the inrush of the fluid into the gap between the wing pairs. This lowers the angle of attack and is thought to create a more attached LEV. The observed differences of the LEV formation associated to the downward flow magnitudes could explain slightly higher force generation of the original wing.

6.4   Flow Visualization in the Wake of the DelFly in Forward Flight

In the previous sections, we limited our investigation and discussion to the DelFly in hovering flight configuration with a particular focus on the force generation, defor- mation and flow structures of the flapping wings since the main objective was to identify the underlying mechanisms behind the flapping-wing flight of the DelFly in a simplified configuration. However, the DelFly does not only hover but also per- forms forward flight. Moreover, in the forward flight configuration, the downstream evolution of flapping wing vortical structures is of vital importance due to possible interaction of these with the tail placed in the wake. Accordingly, this section focuses on the temporal evolution of the flow structures in the wake of the DelFly flapping wings in forward flight configuration. Note that the material presented in this section is published in [21].

One of the few reported attempts to visualize the near wake region of a flapping- wing MAV was made by Ren et al. [23]. They performed PIV measurements in the wake of flapping wings of an MAV that was designed at Wright State University, in chordwise planes at four different spanwise positions. From their observations they concluded that a vortex ring is shed into the wake during the fling phase. They also described the three-dimensional character of the vortex ring based on the inter- pretation of the data from the planar velocity fields acquired at different spanwise locations. Ghosh et al. [9] performed Stereo-PIV measurements in a chordwise plane located in the near wake of a flapping butterfly-shaped wing model which was driven by a 4-bar linkage system. They captured the evolution of flow structures and the spanwise flow at different angles and they reported the dependence of ejected trailing edge vorticity on the flapping frequency.

The wake structure behind the flapping wings of MAVs can be expected to bear similarities to the wake of flapping animal species. The three-dimensional charac- teristics of the wake structure have been investigated in several studies, by their reconstruction from time-resolved planar PIV measurements, using a convection model (Taylors hypothesis). Experiments performed in the wake of different species of birds revealed relatively simple structures: a compact starting vortex at the start of the downstroke and more diffuse stop vortices [14,25]. It was also shown that during the slow flight of birds, most vortical structures are generated during the downstroke whereas the upstroke is almost aerodynamically unemployed. Contrary to birds, experiments on bats suggest a fairly complicated wake structure. Hedenström et al. [13] studied wakes of *G. soricina* bats via the PIV technique. They showed that sepa- rate vortex loops are generated by each wing that consist of the starting vortex, the tip vortex and root vortex, in which the bat wake differs from the bird wake. Moreover, it was observed that, during the end of upstroke at medium and high flight speeds, the outer part of the wing generates a vortex dipole structure with opposite sense to the main vortex. This reversed vortex dipole, which is actually a part of the vortex loop, generates an upwash which in turn produces a negative lift [18]. This negative lift can be regarded as one of the reasons why birds, which retract their wings to make them aerodynamically inactive during the upstroke, outperform bats in terms of span efficiency and flight efficiency [17]. The vortex loops are also observed at smaller length scales. Flow visualizations around and in the near wake of free-flying bumblebees revealed that each wing generates a separate vortex loop that is comprised of LEV, TV and RV with no linkage between the left and right wing structures [4]. On the other hand, PIV measurements in the near-wake of a Hawkmoth (*Manduca Sexta*) exhibited great resemblance to the elliptical vortex loops [3]. Recently, volumetric flow field measurements (Tomographic PIV) were performed in the wake of desert locusts in a volume of $60 \times 80 \times 4 \, \text{mm}^3$ [2]. The three-dimensional wake was visualized in substantial detail, which showed that animal wakes can deform which cannot be detected by use of planar measurement techniques in combination with the convection model, pointing out the importance of volumetric visualization of the wake structures for a correct estimation of lift and thrust.

It is clear that the wake structure of flapping wings, which contains information about the time history of aerodynamic structures and generated forces, requires a detailed investigation, especially for the case of multi-wing MAVs. For example in the specific case of the DelFly, the wake structure is even more compelling as it has two wing pairs and a tail that is presumably interacting with the vortical structures leaving the wings. At these conditions, the size of the wake makes application of Tomographic PIV not feasible [24]. Therefore, in the present investigation the flow in the wake of the flapping wings of the DelFly was captured at several streamwise positions via time-resolved Stereo-PIV measurements. The three-dimensional wake was subsequently reconstructed by use of two different approaches. Firstly, the time-resolved three-component velocity field data of a single measurement plane is convected with the free-stream velocity to yield a spatio-temporal wake reconstruction. Secondly, a Kriging regression technique with a local error estimate was used to generate an instantaneous volumetric representation of the wake structure on a spatial grid that is finer than the spacing between the measurement planes [7]. The main objectives of this wake study are: (1) to investigate the influence of the reduced frequency on the wake structure and (2) to compare the two different approaches, in order to estimate the importance of self-induced wake deformation.

6.4.1 Experimental Setup and Analysis Methods

The experiments were performed in a low-speed wind tunnel at the Aerodynamics Laboratory of TU Delft. A complete DelFly II model was used as the experimental object. The measurements were carried out for free-stream velocities in the range of $2-6 \, \text{m/s}$. The corresponding Reynolds number based on the mean wing chord-length ($\bar{c} = 80 \, \text{mm}$) and U_∞ ranges from 10,000 to 30,000.

The DelFly model was positioned in the test section of the wind tunnel in a forward flight configuration at $0°$ angle of attack and attached to a balance mechanism (Fig. 6.21). The balance mechanism is equipped with two brushed servo motors that allow to change the pitch and yaw angle of the model. The forces and moments were captured by use of a six-component force sensor (ATI Nano-17 with a maximum sensing value of 25 N in x, y and 35 N in z direction with a resolution of

Fig. 6.21 Experimental
setup for the wake
reconstruction study

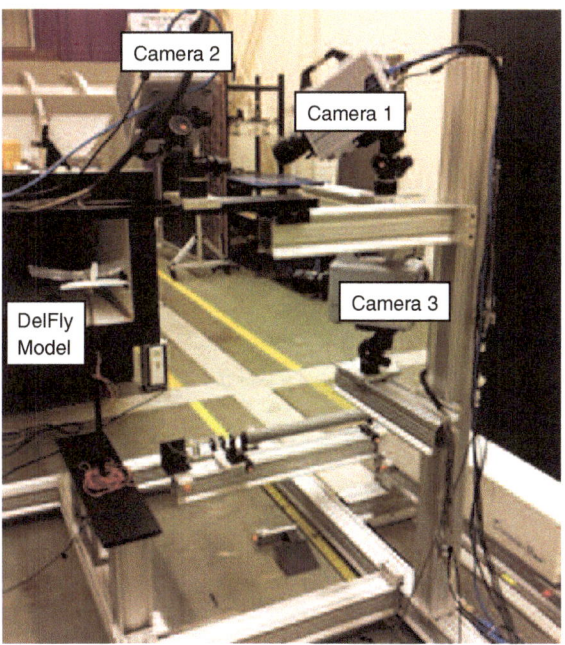

1/160 N) that is attached to the balance mechanism, at a recording rate of 25 kHz. The raw thrust data (viz. X-force in the DelFly body coordinate system) were fil-tered with a Chebyshev Type II low-pass filter with −80 dB attenuation of the stop-band. A forward-backward filtering technique was used in order to prevent time-shift of the data. The cut-off frequency was selected based on the flapping frequency to allow the first two harmonics of the force oscillations to be present in the resultant data according to the findings of the in-air and in-vacuum force measurements.

In addition to the free-stream velocity, also the flapping frequency (f) of the DelFly wings was varied, in the range of 5.7–11 Hz by use of the microcontroller system that was also utilized for the phase determination of the wings and synchro-nization of the PIV acquisition.

High-speed stereo-PIV measurements were performed at 12 consecutive planes with a distance of 10 mm between each other (Fig. 6.22). The first measurement plane was positioned 10 mm downstream from the trailing edge of the wings. It should be noted that the inertial (global) coordinate system was used in the presentation of the wake reconstruction results (see Fig. 6.1a). The measurements were performed only on one side of the wake because of field-of-view size restrictions and imaging blockage by the DelFly structure, however, the wake is assumed to be nominally symmetric with respect to the DelFly fuselage. The wake flow was seeded with a water-glycol based fog o droplets with a mean diameter of 1 μm. The flow was illuminated by a double pulse Nd:YLF high-speed laser (Quantronix Darwin Duo) at a wavelength of 527 nm. The 4 mm thick laser sheet was kept at a fixed position and

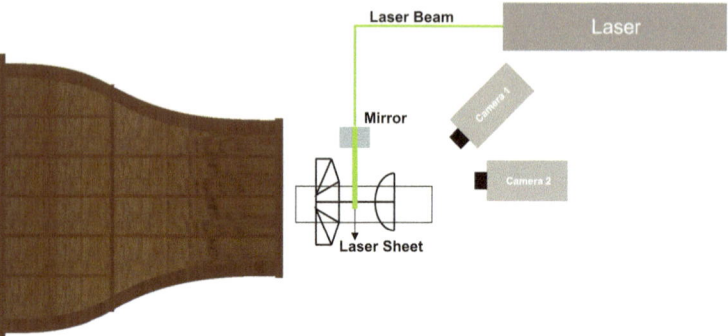

Fig. 6.22 Sketch of the top view of the experimental setup for the wake reconstruction study

the balance mechanism was moved forward or backward to conduct the measurement at a different streamwise position with respect to the DelFly model. Three CMOS cameras with a resolution of 1024×1024 pixels were utilized to capture the images of tracer particles. At each streamwise position, a field of view of $200\,\text{mm} \times 200\,\text{mm}$ was captured at a magnification factor of approximately 0.1 and a digital resolution of 5 pixels/mm. Double-frame images of the tracer particles were recorded for a duration of a second at a recording rate of 250 Hz, which allows for a time-resolved measurement of the flow field, as the flapping frequency ranges between 6 and 12 Hz. The commercial software Davis 8.0 (LaVision) was used in data acquisition, image pre-processing, stereoscopic correlation of the images, and further vector post-processing. As a result, each flow field at different streamwise locations was represented with 5000 vectors with a spacing of 3 mm in each direction.

The microcontroller system was used in order to synchronize the flapping motion with the PIV measurements, which allowed identification of the images that were captured at the same phase of the flapping motion at different measurement planes. For a more detailed explanation of the experimental setup, the reader is referred to Percin et al. [19, 21, 22].

6.4.2 Spatio-Temporal Wake Reconstruction

The spatio-temporal reconstruction was performed for the initial interpretation of the physical phenomenon. For this purpose, time-series measurements in a single measurement plane (i.c., 40 mm downstream of the trailing edge of the wings) were employed to generate three-dimensional representation of the wake structures by using a convection model (Taylors hypothesis). This implies that the data of the measurement plane is translated with the free-stream velocity (with an assumption of non-deforming wake and neglecting the induced velocities), which results in spatial resolution of 8–24 mm in the streamwise direction for the given image recording rate of 250 Hz and for the considered free-stream velocity range (2–6 m/s).

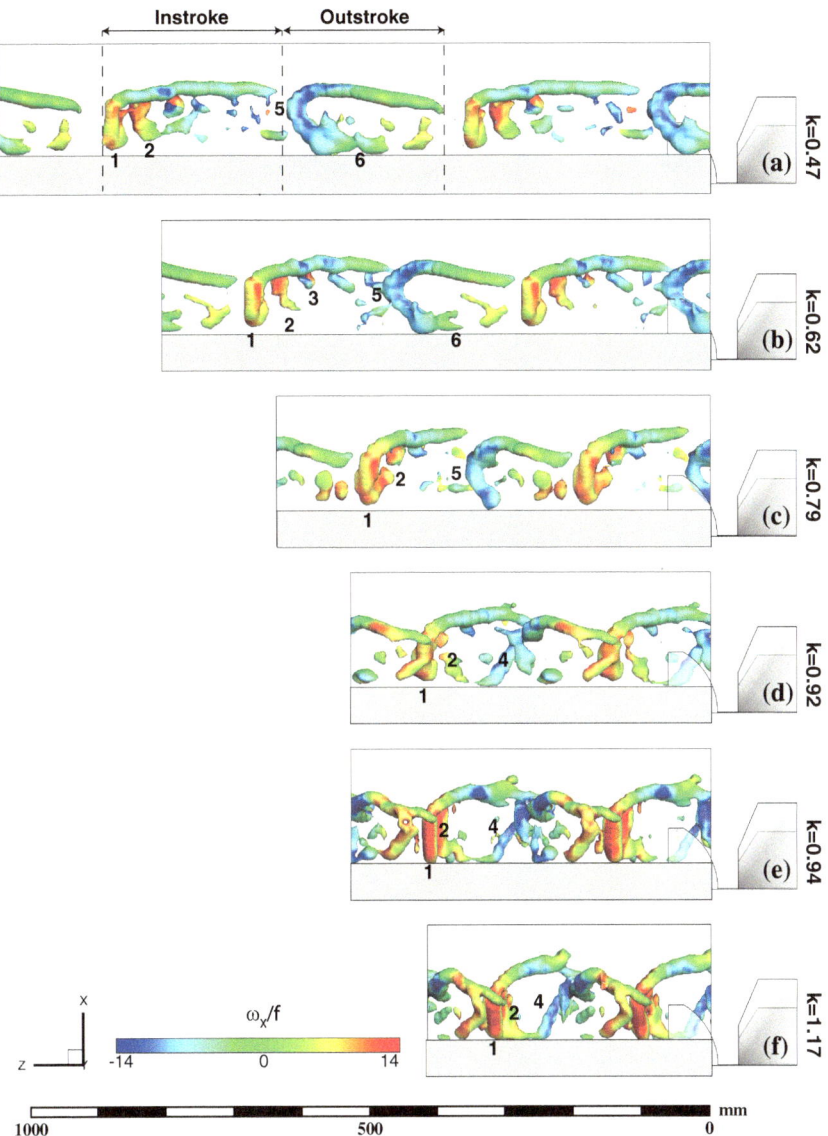

Fig. 6.23 Perspective view of the wake structures visualized by isosurfaces of $Q/f^2 = 10$ and colored by ω_z/f for two periods of the flapping motion (flow is the positive z-direction in the inertial coordinate system) for the case of $k = 0.47$ ($U_\infty = 3$ m/s and $f = 5.7$ Hz)

A perspective view of the wake structures is shown in Fig. 6.23 for the case of reduced frequency of 0.47 (corresponds to a free-stream velocity of $U_\infty = 3$ m/s and flapping frequency of 5.7 Hz). Side and bottom views of the vortical structures for a

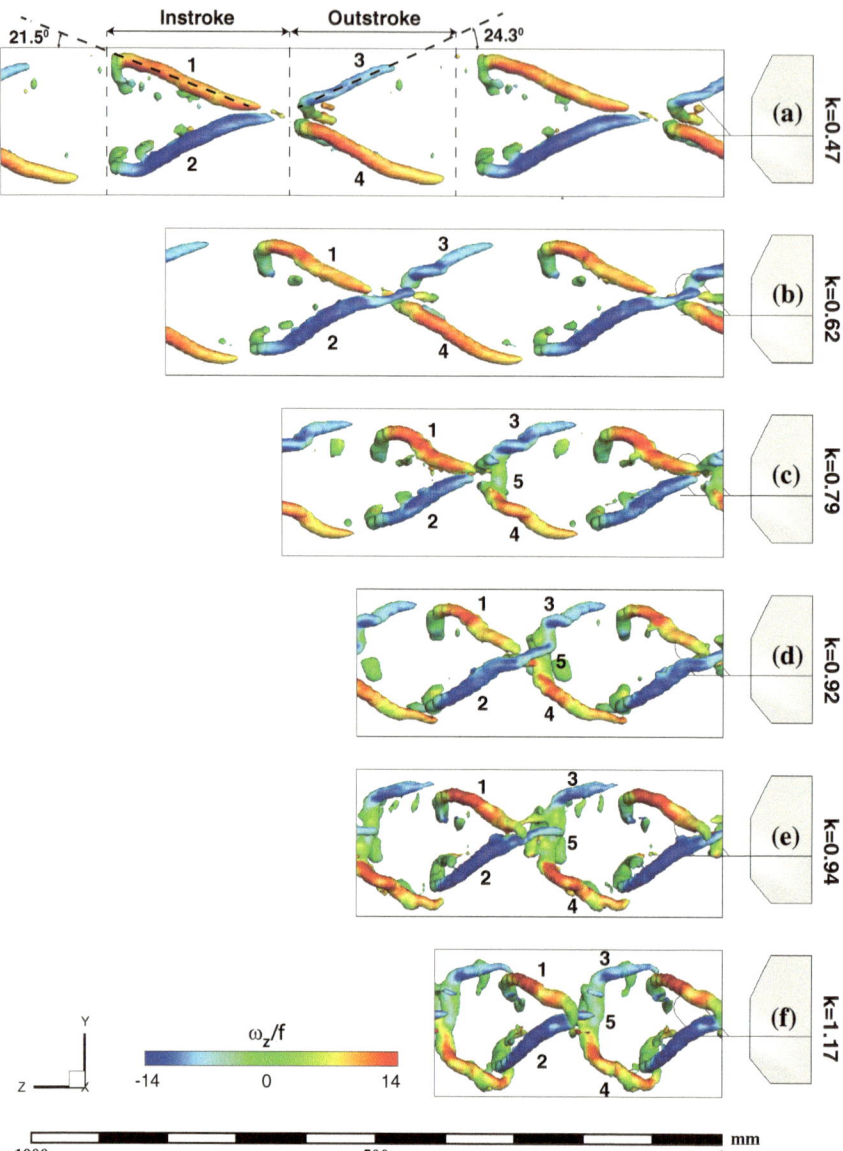

Fig.6.24 Side view of the wake structures visualized by isosurfaces of $Q/f^2 = 15$ and colored by ω_z/f for two periods of the flapping motion (flow is the positive z-direction in the inertial coordinate system). **a** $k = 0.47$ ($U_\infty = 3$ m/s and $f = 5.7$ Hz), **b** $k = 0.62$ ($U_\infty = 3$ m/s and $f = 7.4$ Hz), **c** $k = 0.79$ ($U_\infty = 3$ m/s and $f = 9.4$ Hz), **d** $k = 0.92$ ($U_\infty = 3$ m/s and $f = 11$ Hz), **e** $k = 0.94$ ($U_\infty = 2$ m/s and $f = 7.5$ Hz), **f** $k = 1.17$ ($U_\infty = 2$ m/s and $f = 9.3$ Hz)

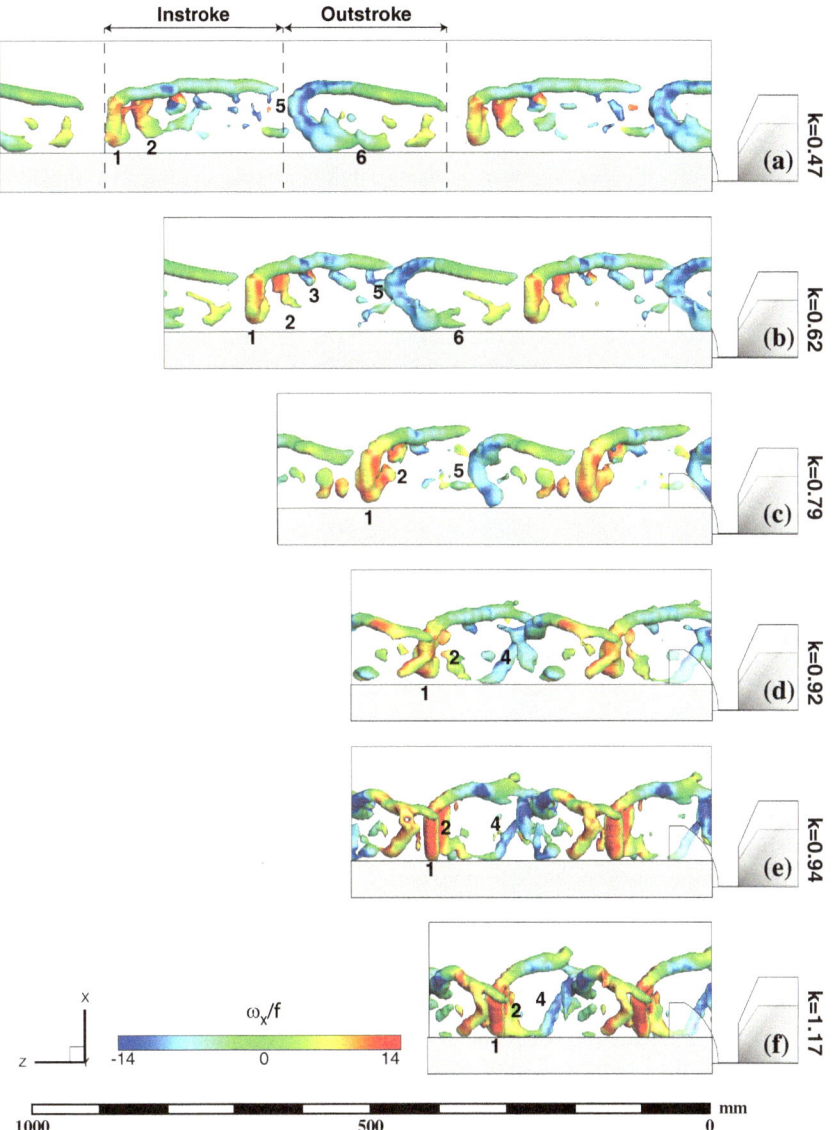

Fig. 6.25 Bottom view of the wake structures visualized by isosurfaces of $Q/f^2 = 8$ and colored by ω_x/f for two periods of the flapping motion (flow is the positive z-direction in the inertial coordinate system). **a** $k = 0.47$ ($U_\infty = 3$ m/s and $f = 5.7$ Hz), **b** $k = 0.62$ ($U_\infty = 3$ m/s and $f = 7.4$ Hz), **c** $k = 0.79$ ($U_\infty = 3$ m/s and $f = 9.4$ Hz), **d** $k = 0.92$ ($U_\infty = 3$ m/s and $f = 11$ Hz), **e** $k = 0.94$ ($U_\infty = 2$ m/s and $f = 7.5$ Hz), **f** $k = 1.17$ ($U_\infty = 2$ m/s and $f = 9.3$ Hz)

selection of parameters with corresponding reduced frequencies ranging from 0.47 to 1.17 are shown in Figs. 6.24 and 6.25 (cases a to d correspond to a free-stream velocity of $U_\infty = 3$ m/s and flapping frequency of 5.7, 7.4, 9.4, and 11 Hz, respectively; e and f are for $U_\infty = 2$ m/s and 7.5 and 9.3 Hz, respectively). Two periods of the flapping motion are depicted, visualized by means of isosurfaces of the Q criterion [15] which is non-dimensionalized by the square of the flapping frequency. The isosurfaces of dimensionless Q criterion (Q/f^2) are colored by the dimensionless vorticity component parallel with the free-stream (ω_z/f) in the side and perspective views and by the dimensionless spanwise vorticity component (ω_x/f) in the bottom views. Note that the region in the vicinity of the tail is masked during PIV processing due to intensive reflections underneath the tail and lack of illumination above it. Additionally, to assist the interpretation of the results, the DelFly wings are schematically displayed in the images as an indication of the flapping phase (particularly in the case of the spatial wake reconstruction), however, they are represented here as rigid bodies whereas in reality there is a significant amount of deformation of the wings during the flapping motion.

It can be observed that the wake of the flapping wings is composed of well-organized vortical structures and is dominated by the counter-rotating tip vortices of upper and lower wings (Fig. 6.23). The instroke starts with the formation of starting vortices (Fig. 6.23; 1 and 2) linked to the tip vortices both upper and lower wings (Fig. 6.23; 3 and 4). During the outstroke, similarly the wings shed coherent tip vortices (Fig. 6.23; 5 and 6) of the opposite sense. For the lowest reduced frequency the vortical structures of the instroke and outstroke phases are clearly separated (Fig. 6.24a), indicating that little vortex activity occurs during the stroke reversal moments. During the instroke phase of the flapping motion, both the upper and the lower wing form symmetrical tip vortices (Fig. 6.24a–e; 1 and 2, respectively) about an axis situated slightly above the centerline of the DelFly (i.e. the fuselage of the DelFly which is depicted as a horizontal line in the DelFly drawings in Fig. 6.24) due to the 12° dihedral angle of the DelFly wings. Tip vortices shed during the outstroke also show a similar behavior in terms of symmetry in general (Fig. 6.24a–e; 3 and 4); however, for the lowest reduced frequencies (Fig. 6.24a, b), relatively weaker vortical wake structures occur. Also the inclination of the tip vortices differs, as indicated in the figure, being larger for the outstroke than for the instroke. These difference between the vortical structures of in- and outstroke can be explained by the different wing kinematics during the in-and-outstroke, which is presumably a direct result of the wing-wing interaction occurring at the onset of the outstroke (viz. clap-and-peel) together with the asymmetry in the flapping motion profile (Fig. 6.2).

The most obvious effect of the reduced frequency reflected in the visualizations of Figs. 6.24 and 6.25 is the wave length of the wake being inversely proportional to k. This is a logical outcome, as the reduced frequency is, by its definition, proportional to the ratio of the wavelength of the vortex wake (U_∞/f) to the characteristic length of the flapping wings ($\bar{c}$).

Another prominent difference between the considered cases of reduced frequencies is the time shift in the appearance of vortical structures in the reconstructed wakes. The Hall sensor input is used for zero-referencing between the measurements, thus all wake reconstructions are synchronized with respect to the stroke angle of the flapping wings. Therefore, the phase lag suggests different formation and convection times of the wake structures for different reduced frequencies. This implies that the convection time of the vortices from the vicinity of the flapping wings to the measurement plane that is used for the spatio-temporal reconstruction is not only determined by the free-stream velocity but might also be affected by these structures own induced velocities as well as flapping-wing induced velocities (e.g., the momentum jet during the clap phase). The formation time of these structures, on the other hand, is a more complex and difficult-to-estimate parameter as it might depend on a number of interrelated variables (fluid-deforming structure interactions, wing-wing interactions, wing kinematics, fluid dynamics effects, etc.). In this regard, earlier flow visualization studies around the flapping wings of the DelFly [12] indeed reported different TEV developments at different flapping frequencies. It was shown that increasing flapping frequency delays the formation of the TEV as well as increasing its circulation.

Therefore, the observed phase lag can be explained by the formation time of vortical structures that changes with varying flapping frequency and finds its likely origin in wing deformation characteristics. The structural dynamics of equivalent-DelFly-wings have analyzed by use of a simplified analytical model [20] which showed that the wing-inertia term has a considerable effect on the resultant wing deformation especially at the stroke reversals, as discussed previously. The passive behavior of the flexible wing surface and stiffeners alters significantly due to wing inertia. At stroke reversal, the wing body continues moving with its inertia and changes direction in motion later than the leading edge spars. The delay and amount of deformation are enhanced by the flapping frequency and free-stream velocity

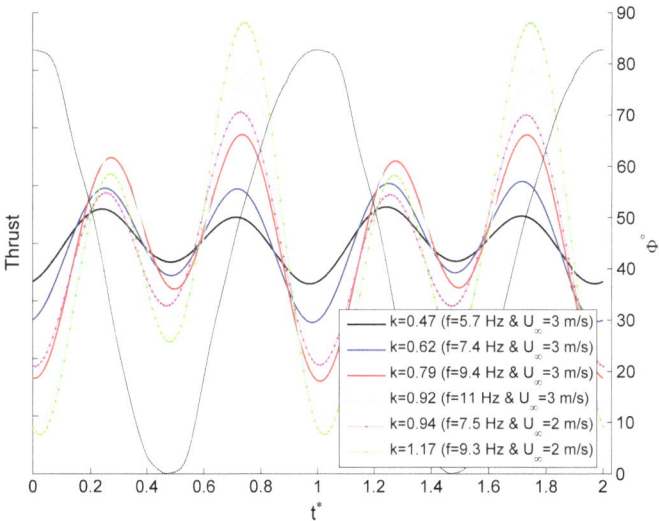

Fig. 6.26 Time variation of thrust plotted for two periods of the flapping motion for different reduced frequencies (mean force is subtracted from the measured forces for a better comparison) complemented with the variation of stroke angle (*black dashed line*)

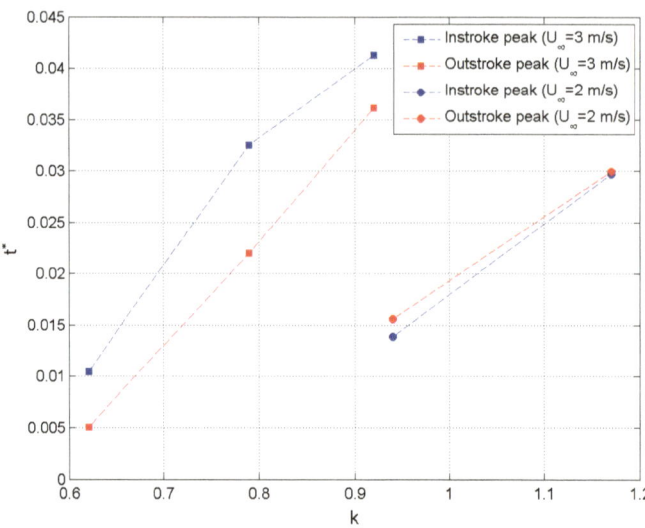

Fig. 6.27 Relative time shift of the maximum force generation instants of the instroke and the outstroke with respect to that of the case $k = 0.47$

which results in the formation of vortical structures with a time shift. In this context, comparison of thrust profiles (Fig. 6.26) also displays a lag in non-dimensional times when the peak forces are generated. Figure 6.27 displays the relative time delays with respect to the lowest reduced frequency case. It is clear that during both instroke and outstroke phases, generation of maximum forces delays with increasing k at a constant U_∞ (viz., increasing f) and smaller lag is observed at lower U_∞ cases.

A distinct feature that comes out of the comparison of the spatio-temporal wake reconstructions is the changing interaction characteristics between the vortical structures. The interaction takes place between the formations of the instroke and outstroke phases as well as vortical structures of the upper and lower wings within a single stroke. The former emerges for the cases of k greater than 0.47 (Fig. 6.24b–f). For these cases, tip vortices of the instroke (1 and 2) and the outstroke (3 and 4), respectively, are situated close to each other and co-rotating vortices of these strokes even interact with each other. On the other hand, the lowest k case (Fig. 6.24a) represents individual structures of different strokes with no interaction observed with the current Q isovalue. Furthermore, gaps between the consecutive strokes with no prominent wake structures can be observed for this case. The other type of interaction mostly occurs during the outstroke as the wing-wing interaction takes place at the onset of this phase. For the cases of k greater than 0.62, tip vortices of the upper and the lower wings are linked to each other and form an arrow-shaped vortex loop (Fig. 6.24c–f; 5) whereas for the two lowest k cases, this interaction is absent or only weak. The presence of such a vortex formation is also correlated with the thrust generation mechanism of the flapping motion (Fig. 6.26). For the lowest k case, higher thrust is produced during the instroke than for the outstroke (relative difference of

28 %). The relative strength of the instroke and outstroke vortical structures is also in accordance with this behavior. This result, on the other hand, contradicts the previous studies on the hovering DelFly [6,12], which reported larger force generation during the outstroke thanks to the enhancing effect of the peel motion. Equivalent thrust is generated for the case of $k = 0.62$ during both the instroke and the outstroke. This contradiction vanishes at larger values of k, where considerably greater thrust is generated during the outstroke with footprints of wing-wing interaction in the wake. Apparently the clap-and-peel mechanism that is prominent in hovering flight, is active in the forward flight condition only for a sufficiently high value of the flapping frequency, but becomes deactivated at low values of the reduced frequency (low frequency or high free stream velocity).

Comparison of the transverse wake structures (viz. leading and trailing edge vortices) exhibits more complex interaction characteristics as shown in Fig. 6.25. Note that the region where there is no reliable data due to reflection and shadow of the tail is covered with a gray band and only the wake structures of the lower wing are visualized for ease of comparison. Common to all cases, the instroke starts with the formation of starting vortices (Fig. 6.25a–f; 1) of both wings. The starting vortex is then followed by the shedding of secondary trailing edge vortex (2). There is a weak trace of leading edge vorticity for the case of $k = 0.62$ (Fig. 6.25b; 3). The LEV of the instroke sheds at the end of the instroke and appears as a prominent structure in the wake as a part of a vortex loop with the root vortex which looms for the cases of k greater than 0.79 (Fig. 6.25d–f; 4). In the other cases, presumably it sheds and interacts with the starting vortex of the outstroke and appears as a single structure in the wake. A starting vortex forms at the beginning of the outstroke clearly for the cases of $k = 0.47 - 0.79$ (Fig. 6.25a–c; 5) contrary to the theory of Weis-Fogh [28] which postulates annihilation or absence of the starting vortex at the onset of the outstroke due to opposite bound circulations of the wings. Moreover, there are indications of root vortices forming during the outstroke for the two lowest k cases which are linked to the starting vortices (Fig. 6.25a, b; 6). On the other hand, less coherent trailing edge formations are present for the higher k values mostly closer to the wing tips at the onset of the outstroke which is intuitive as the outer parts of the wings move away initially. However, for these cases there is no sign of a root vortex of the outstroke interacting with the starting vortex. Instead, as shown earlier, vortices of upper and lower wings interact and form a vortex loop.

6.4.3 Spatial Wake Reconstruction

For a spatial wake reconstruction, a Kriging regression technique was applied to the flow fields from consecutive measurement planes in order to reconstruct a true spatial representation of the instantaneous three-dimensional wake of the flapping DelFly wings, for a selected phase in the flapping cycle [7]. In this procedure two equidistant planes are interpolated between each of the measurement planes resulting in a spatial resolution of 3.33 mm in the streamwise direction (which is then comparable to the in-plane resolution). The implemented Kriging technique can be regarded as an improved version of the ordinary Kriging interpolation, in that it is modified to take into account the local measurement uncertainty in the regression process. A low-fidelity statistical model that is based on the signal to noise ratio (SNR) of the PIV measurements was developed to estimate the local error in the velocity fields. The error model is then incorporated in the Kriging algorithm to diminish the effect of measurement points with high uncertainty on the interpolated points.

Spatial reconstruction results are presented in this section for a selection of reduced frequencies (i.e., $k = 0.47, 0.62,$ and 0.92) and phases of the flapping motion. The phases have been selected so as to capture the most prominent vortical structures of the strokes in the field of view.

Figure 6.28 displays the three-dimensional wake formations in the case of a reduced frequency of $k = 0.47$ ($U_\infty = 3$ m/s and $f = 5.7$ Hz) at a time instant close to the end of the instroke ($t^* = 0.335$ and $\Phi = 15.5°$), when most of the

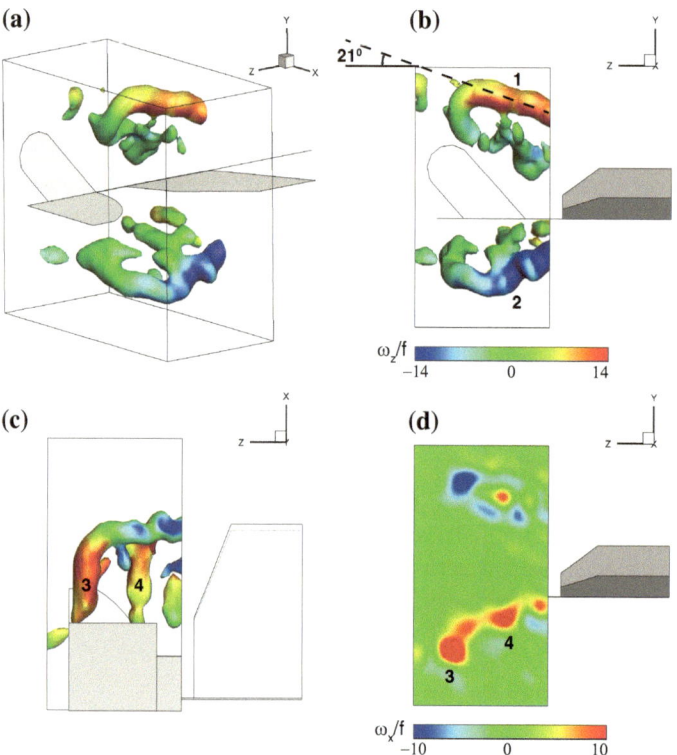

Fig. 6.28 Wake structures for the case of $k = 0.47$ ($U_\infty = 3$ m/s and $f = 5.7$ Hz) at $t^* = 0.335$ ($\Phi = 15.5°$). **a** Perspective view of the isosurfaces of $Q/f^2 = 10$ colored by ω_z/f. **b** *Side view* of the isosurfaces of Q/f^2 colored by ω_z/f. **c** *Bottom view* of the isosurfaces of Q/f^2 colored by ω_x/f (note that only lower half of the wake is visualized). **d** Contours of the out-of-plane vorticity (ω_x/f) in an x-plane positioned at 70 % of the span length

vortical structures generated during the instroke appear in the observed region of the wake. In general, the reconstructed wake displays similar characteristics as the spatio-temporal reconstruction of the wake (Figs. 6.24a, 6.25a): no prominent vortex activity between the outstroke and the instroke phases (Fig. 6.28b, c); no interaction between the formations of the upper and lower wings; a symmetrical tip vortex formation about the dihedral axis (Fig. 6.28b; 1 and 2). Furthermore, the inclination angle of the upper wings tip vortex tube for the spatio-temporal reconstruction and the spatial reconstruction show a good agreement (see Figs. 6.24a, 6.28, 6.29b). This similarity validates the assumptions underlying the spatio-temporal wake reconstruction (i.e., non-deforming wake and free-stream velocity as the convection velocity) for this particular case.

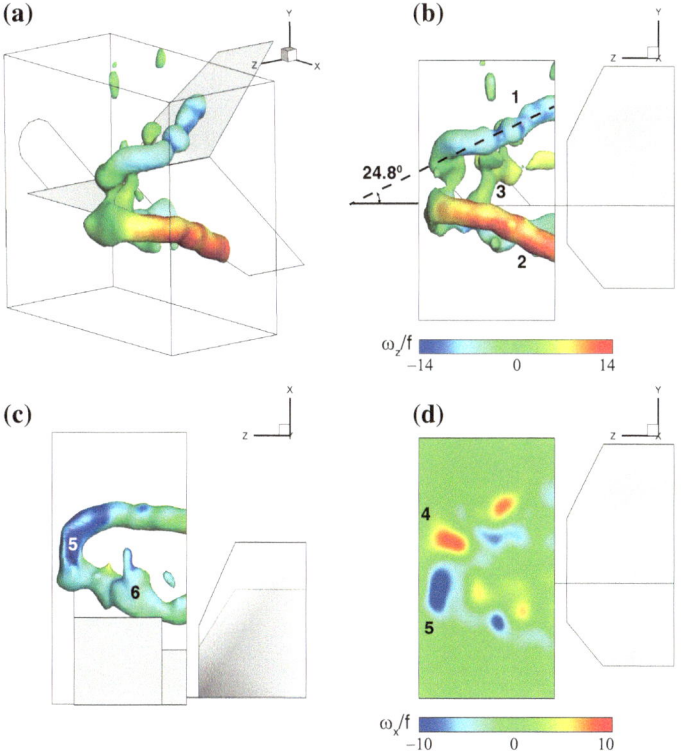

Fig. 6.29 Wake structures for the case of $k = 0.47$ ($U_\infty = 3$ m/s and $f = 5.7$ Hz) at $t^* = 0.91$ ($\Phi = 79°$). **a** Perspective view of the isosurfaces of $Q/f^2 = 10$ colored by ω_z/f. **b** Side view of the isosurfaces of Q/f^2 colored by ω_z/f. **c** *Bottom view* of the isosurfaces of Q/f^2 colored by ω_x/f (note that only lower half of the wake is visualized). **d** Contours of the out-of-plane vorticity (ω_x/f) in an x-plane positioned at 70 % of the span length

Examination of transverse vortical structures also indicates similar behavior with the spatio-temporal wake reconstruction results, but with more details thanks to a better resolution in the streamwise direction and more information in the region closer to the wing roots provided by the upstream measurement planes. The bottom view of the three-dimensional vortical structures (Fig. 6.28c) is complemented with the contours of the out-of-plane vorticity in an x-plane that is positioned at 70 % of the span length (Fig. 6.28d). It should be noted that this section cut is not normal to the chordwise oriented vortical structures and only used for visualization purposes. The instroke starts with the shedding of a clockwise rotating starting vortex for the lower wing (Fig. 6.28c, d; 3) that is then followed by continuous trailing edge vorticity and secondary trailing edge vortex shedding (Fig. 6.28c, d; 4). The upper wing displays an identical vortex shedding mechanism, although it is not properly visualized in Fig. 6.28d because the structures of the upper wing have a larger angle with the x-plane with respect to that of the lower wing for this phase.

In Fig. 6.29, the three-dimensional wake is depicted approximately at the end of the outstroke ($t^* = 0.91$ and $\Phi = 79°$). At this time instant, vortex structures formed during the outstroke appear in the field of view. As discussed earlier in the spatio-temporal reconstruction results, tip vortices of upper and lower wings are present in the wake (Fig. 6.29b; 1 and 2) linked to the starting vortices of the outstroke (Fig. 6.29d; 4 and 5). Similar to the structure shown in Fig. 6.25a; the starting vortex and the root vortex of the outstroke phase are present in the wake (Fig. 6.29c; 5 and 6). No interaction takes place between the tip vortices of upper and lower wings, but there is a sign of interaction between the root vortices of both wings (Fig. 6.29b; 3).

In Fig. 6.30, the wakes of the flapping wings are compared for different reduced frequencies ($k = 0.47$, 0.62, and 0.92) at the same phase of the flapping motion ($t^* = 0.445$ and $\Phi = 15.5°$). The comparison reveals and confirms two features of the wake of the flapping wings that are assessed as a result of spatio-temporal

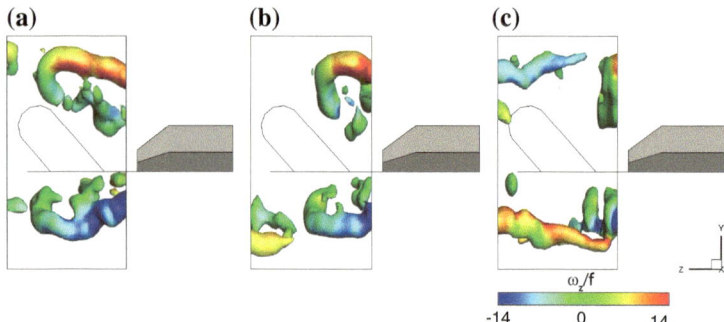

Fig. 6.30 Side view of wake structures visualized by isosurfaces of $Q/f^2 = 6$ and colored by ω_z/f (flow is the positive z direction) at $t^* = 0.335$ ($\Phi = 15.5°$). **a** $k = 0.47$ ($U_\infty = 3$ m/s and $f = 5.7$ Hz), **b** $k = 0.62$ ($U_\infty = 3$ m/s and $f = 7.4$ Hz), **c** $k = 0.92$ ($U_\infty = 3$ m/s and $f = 11$ Hz)

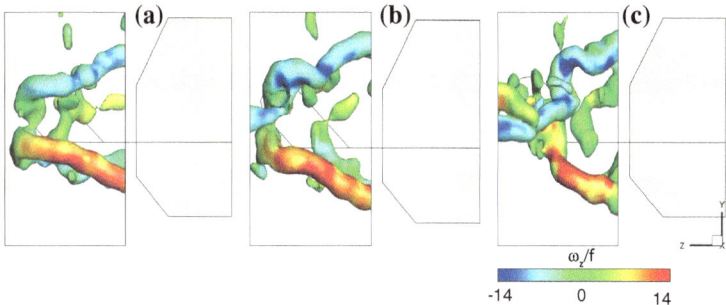

Fig. 6.31 *Side view* of wake structures visualized by isosurfaces of $Q/f^2 = 6$ and colored by ω_z/f (flow is the positive z direction). **a** $k = 0.47$ ($U_\infty = 3$ m/s and $f = 5.7$ Hz) at $t^* = 0.91$ ($\Phi = 79°$), **b** $k = 0.62$ ($U_\infty = 3$ m/s and $f = 7.4$ Hz) at $t^* = 1$ ($\Phi = 82°$), **c** $k = 0.92$ ($U_\infty = 3$ m/s and $f = 11$ Hz) at $t^* = 1.08$ ($\Phi = 79°$)

wake analysis. Firstly, there is the phase lag in the appearance of the structures - for the lowest flapping frequency case (Fig. 6.30a), the vortices of the instroke are already formed and convected more than half of the reconstruction length in the streamwise direction, whereas for the highest flapping frequency case, the tip vortices of the previous downstroke are still in the wake and starting vortices of the instroke newly form (Fig. 6.30c). Secondly, changing interaction characteristics between the formations of consecutive strokes are observed. In the view of the latter, no prominent vortex activity during the stroke reversal case at the lowest reduced frequency case evolves into a more complex vortex interaction with increasing flapping frequency.

To compensate for the phase lag to achieve a clearer comparison of the outstroke wake structures, slightly shifted phases of the flapping cycle were reconstructed for increasing flapping frequency (Fig. 6.31). The comparison of the outstroke structures shows the evolution of wake structures from non-interacting upper and lower wing case to fully interacting cases. Moreover, it is clear that the structures of the instroke and the outstroke have a linkage in between for the higher flapping frequency cases (Fig. 6.31b, c). In addition to the arrow-shaped vortex loop that is formed as a result of clap-and-fling motion being dominant for the highest flapping frequency case, there is a less coherent structure present formed by the root vortices.

6.5   Conclusions

In this chapter, we gave an overview of experimental studies focusing on the aerodynamics of the DelFly flapping-wing flight. In the first part, results of the experiments that were performed in-air and in-vacuum conditions are presented. It is shown that the first two harmonics in the X-force (i.e., lift in hovering configuration) spectrum are certainly associated to the aerodynamic forces. Comparison of the power consumption in the two measurement conditions shows that 64—74 % of the total consumption is due to aerodynamic effects, while the contribution due to inertial and elastic effects reaches maximum 20 % of the total consumption. Further inspection on the X-force oscillations in combination with the high-speed images clarifies that the first (fundamental) harmonic is related to the clap-and-peel mechanism that takes place once during a flapping cycle, whereas the second harmonic is likely to originate from the leading edge vortex (delayed stall) mechanism, which occurs twice both during the instroke and the outstroke. It is shown that mean X-force generated during the outstroke is 75 % more than that of the instroke due to enhancing effect of clap-and-peel motion. Power consumption in a wing-beat cycle shows a similar variation with the X-force generation yet with a phase lag.

As a result of the systematic wing optimization study, a wing layout with an improved aerodynamic performance was attained. The improved wing design shares the same wing shape, surface area, span length and surface thickness with the original (old) wing design except the orientation and positioning of the wing stiffeners. It was shown that use of improved wing configuration increases the X-force to power consumption ratio by approximately 5 % although it generates almost the same force with the old wing. This is attributed to the changing stiffness distribution such that the improved wing is more stiff toward the wing tip. Hypothetically. enhanced stiffness in the outwards part of the wings results in more favorable conditions in terms of reducing the resistance forces particularly in the region close to the wing tips where tangential sweeping velocity is higher. The wing deformation measurements at $0.71R$ spanwise location also revealed that the original wing configuration display more flexible characteristics and has an irregular shape particularly during stroke reversals. Experimental flow field measurements by use of stereo-PIV technique shows the effects of varying stiffness characteristics on the vortical structures for the original and improved wing designs. A larger LEV that is rather detached from the wing surface forms in the case of original wing configuration, whereas the improved wing features an LEV that is spread and attached on the wing surface.

In the last part, unsteady three-dimensional wake structures generated by the flapping wings of the DelFly II in forward flight condition are explored. The three-dimensional wake configuration was reconstructed from the planar PIV measurements by two approaches: (1) a spatiotemporal wake reconstruction obtained by convecting the time-resolved, three-component velocity field data of a single measurement plane with the free-stream velocity; (2) for selected phases in the flapping cycle a direct three-dimensional spatial wake reconstruction is interpolated form the data of the different measurement planes, using a Kriging regression technique. Comparison of spatiotemporal reconstruction of the wakes for different reduced frequencies revealed the effects of the flapping frequency on the wake formation and force generation mechanisms. First, various interaction forms are observed and described in the compared cases. For the lowest reduced frequency case ($k = 0.47$), there is no interaction observed between the wake structures of the consecutive strokes and there are gaps with no vortex activities in between. The situation changes with increasing reduced frequency with the appearance of interacting and coupled structures of the successive strokes. A more important interaction takes place between the upper and lower wings at the start of the outstroke (i.e., clap-and-peel motion) that affects the thrust generation mechanism significantly. Analysis of the results reveals that this mechanism is inactive at the reduced frequencies smaller than 0.62, which reflects in a higher thrust generation during the instroke phase of the flapping motion, in contradiction to what is generally reported for the clap-and-fling phenomenon in hovering flight condition [12, 16], and individual vortex formations of the upper and lower wings. For the reduced frequencies greater than 0.62, however, this wing inter-action mechanism becomes more dominant and generates greater thrust during the outstroke in combination with the arrow-shaped vortex loop that is formed by the tip vortices of both wings. Secondly, increasing flapping frequency for a given free-stream velocity introduces a time shift in the appearance of vortical structures in the wake that is also correlated with the lag in the generation of the maximum thrust during both the instroke and the outstroke. This time shift is associated with the changing wing deformation characteristics, especially at the stroke reversals due to significant contribution of the wing inertia. Comparing the results derived from both methods in terms of the behavior of the wake formations, their phase and orientation indicate that the spatiotemporal reconstruction method allows to characterize the general three-dimensional structure of the wake, but that the spatial reconstruction method can reveal more details due to higher streamwise resolution.

Although only some of the experimental studies are reported in this book, research on the (DelFly) flapping-wing aerodynamics is an ongoing process carried out by the aerodynamics research group. The studies that are being performed by the group ranges from numerical simulations of (flexible) flapping wings [10], Tay et al. [26, 27] to experimental research on the DelFly Micro [8] and to the fundamental research in flapping-wing aerodynamics in simplified configurations Percin et al.[20, 29].

References

1. R.J. Adrian, K.T. Christensen, Z.C. Liu, Analysis and interpretation of instantaneous turbulent velocity fields. Exp. Fluids **29**(3), 275–290 (2000)
2. R.J. Bomphrey, P. Henningsson, D. Michaelis, D. Hollis, Tomographic particle image velocime-try of desert locust wakes: instantaneous volumes combine to reveal hidden vortex elements and rapid wake deformation. J. R. Soc. Interface/R. Soc. **9**(77), 86–3378 (2012)
3. R.J. Bomphrey, N.J. Lawson, G.K. Taylor, A.L.R. Thomas, Application of digital particle image velocimetry to insect aerodynamics: measurement of the leading-edge vortex and near wake of a Hawkmoth. Exp. Fluids **40**(4), 546–554 (2006)
4. R.J. Bomphrey, G.K. Taylor, A.L.R. Thomas, Smoke visualization of free-flying bumblebees indicates independent leading-edge vortices on each wing pair. Exp. Fluids **46**(5):811–821 (April 2009)

5. B. Bruggeman, Improving flight performance of DelFly II in hover by improving wing design and driving mechanism. Master's thesis, Faculty of Aerospace Engineering, TU Delft, The Netherlands (2010)

6. K.M.E. De Clercq, R. de Kat, B. Remes, B.W. van Oudheusden, H. Bijl, Aerodynamic experiments on DelFly II: unsteady lift enhancement. Int. J. Micro Air Veh. **1**(4), 255–262 (2009)

7. J.H.S. de Baar, M. Percin, R.P. Dwight, B.W. Oudheusden, H. Bijl, Kriging regression of PIV data using a local error estimate. Exp. Fluids **55**(1), 1650 (2014)

8. S. Deng, M. Percin, B. van Oudheusden, B. Remes, H. Bijl, Experimental investigation on the aerodynamics of a bio-inspired flexible flapping wing micro air vehicle. Int. J. Micro Air Veh. **6**(2), 105–116 (2014)

9. S.K. Ghosh, C.L. Dora, D. Das, Unsteady wake characteristics of a flapping wing through 3D TR-PIV. J. Aerosp. Eng. **25**, 547–558 (2012)

10. T. Gillebaart, Influence of flexibility on the clap and peel movement of the DelFly II. Master's thesis, Delft University of Technology (2011)

11. M. Groen, PIV and force measurements on the flapping-wing MAV DelFly II. Master's thesis, Delft University of Technology (2010)

12. M.A. Groen, B. Bruggeman, B.D.W. Remes, R. Ruijsink, B.W. van Oudheusden, H. Bijl, Improving flight performance of the flapping wing mav delfly ii, in *International Micro Air Vehicle conference Braunschweig, Germany* (2010)

13. A. Hedenstrom, L.C. Johansson, M. Wolf, R. von Busse, Y Winter, G.R. Spedding. Bat flight generates complex aerodynamic tracks. Science (New York, N.Y.) **316**(5826):894–897 (May 2007)

14. H. Anders, M Rosén, G.R. Spedding, Vortex wakes generated by robins Erithacus rubecula during free flight in a wind tunnel. J. R. Soc. Interface/R. Soc. **3**(7):263–76 (April 2006)

15. J. Jeong, F. Hussain, On the identification of a vortex. J. Fluid Mech. **285**, 69–94 (1995)

16. L. Fritz-Olaf, S.P. Sane, M.H. Dickinson, The aerodynamic effects of wing-wing interaction in flapping insect wings. J. Exp. Biol. **208**(Pt 16):3075–3092 (August 2005)

17. F.T. Muijres, L.C. Johansson, M.S. Bowlin, Y. Winter, A. Hedenström. Comparing aerodynamic efficiency in birds and bats suggests better flight performance in birds. *PloS one***7**(5) (January 2012)

18. F.T. Muijres, G.R. Spedding, Y. Winter, A. Hedenström, Actuator disk model and span efficiency of flapping flight in bats based on time-resolved PIV measurements. Exp. Fluids **51**(2), 511–525 (2011)

19. M. Percin, H.E. Eisma, J.H.S. de Baar, B.W. van Oudheusden, B. Remes, R. Ruijsink, C. de Wagter, Wake reconstruction of flapping-wing MAV DelFly II in forward flight, in *International Micro Air Vehicle Conference and Flight Competition* (2012)

20. M. Percin, Y. Hu, B.W. Van Oudheusden, B. Remes, F. Scarano, Wing flexibility effects in clap-and-fling. Int. J. Micro Air Veh. **3**(4), 217–228 (2011)

21. M. Percin, B.W. van Oudheusden, H.E. Eisma, B.D.W. Remes, Three-dimensional vortex wake structure of a flapping-wing micro aerial vehicle in forward flight configuration. Exp. in Fluids **55**(9):1806 (August 2014)

22. M. Percin, H. Eisma, B. Van Oudheusden, B. Remes, R. Ruijsink, C. De Wagter. Flow visualization in the wake of flapping-wing mav delfly iiin forward flight, in *AIAA Fluid Dynamics and Co-located Conferences and Exhibit New Orleans* (2012)

23. H. Ren, Y. Wu, P.G. Huang, Visualization and characterization of near-wake flow fields of a flapping-wing micro air vehicle using PIV. J. Visual. **16**(1), 75–83 (2012)

24. F. Scarano, Tomographic PIV: principles and practice. Measur. Sci. Technol. **24**(1), 012001 (2013)

25. G.R. Spedding, M. Rosén, Anders Hedenström, A family of vortex wakes generated by a thrush nightingale in free flight in a wind tunnel over its entire natural range of flight speeds. J. Exp. Biol. **206**(14), 2313–2344 (2003)

26. W.B. Tay, B.W. Van Oudheusden, H. Bijl, Numerical simulation of x-wing type biplane flapping wings in 3D using the immersed boundary method. Bioinspir. Biomimet. **9**(3), 036001 (2014)
27. W.B. Tay, B.W. van Oudheusden, H. Bijl, Numerical simulation of a flapping four-wing micro-aerial vehicle. J. Fluids Struct. **55**, 237–261 (2015)
28. T. Weis-Fogh, Quick estimates of flight fitness in hovering animals, including novel mechanisms for lift production. J. Exp. Biol. **59**, 169–230 (1973)
29. M. Percin, B.W. van Oudheusden, Three-dimensional flow structures and unsteady forces on pitching and surging revolving flat plates. Exp. in Fluids **56**(2), 1–19 (2015)

Part III
Autonomous Flight

Introduction to Autonomous Flight

7

Abstract

This chapter sets the stage for the research on autonomous flight of the DelFly. First, a general introduction is given to artificial intelligence for robotics. This will permit the layman to understand some of the major challenges in creating autonomous robots, and the different solution approaches. Subsequently, the particular tasks and approaches to autonomous flight of Micro Air Vehicles are discussed. Whereas mainstream approaches are now being applied to ∼1 kg MAVs such as quadrotors, it will become clear that light-weight flapping wing MAVs are best served by a computationally efficient, vision-based approach to autonomous flight. In particular, in the DelFly project we have adopted a purposive vision approach to autonomous flight, in which only the information necessary for the robot's task is extracted. The main goal of the approach is to allow the DelFly to autonomously explore unknown indoor environments, to which end we complement optical flow with other, appearance based, visual inputs.

7.1 Background Information on Artificial Intelligence for Robotics

In this section, we give a brief overview of the history of Artificial Intelligence (AI) in the context of robotics. The overview is not meant to be exhaustive, but aims to provide readers from different disciplines a sufficient background to understand the main problems in the research on autonomous robots and the main approaches followed to solve them.

A central concept in computer science and artificial intelligence is that of an *algorithm*: a list of well-defined instructions that lead to the computation of a function. The instructions should be well-defined in the sense that a computing device can execute them. Simply put, an algorithm is a sort of recipe for computers, telling them what they should do and in what order. In the end, the instructions in an algorithm come down to basic operations such as the multiplication of two numbers or the setting of a memory location. Knowing this, one can imagine that it is easier to make algorithms for calculating the first 100 numbers of a Fibonacci sequence than for recognizing faces in digital images. Namely, in the case of the Fibonacci sequence, we immediately see the basic instructions that have to be executed by the computer: keeping numbers in memory and adding them. An algorithm for recognizing faces is much harder to make. Although even babies can recognize faces, it is difficult for us to translate this capability in a list of instructions that can be executed by the computer. A main reason for this difficulty is that the recognition of faces involves implicit knowledge.

It may then come at no surprise that at the start of the field of artificial intelligence the focus was on closed, abstract domains, in which the knowledge could be made explicit. Alan Turing used computers for cryptography, which allowed to crack the ENIGMA codes in World War II [42]. In addition, he was one of the first researchers to write chess programs. In 1955 Simon and Newell investigated the way in which human subjects tend to solve problems. They made a program that relied on logic reasoning to mimic the problem-solving skills of human beings. Their 'Logic Theorist' proved 38 out of the first 52 theorems in Whitehead and Russell's 'Principia Mathematica' [44], for some theorems even finding more elegant proofs than were known before. The Logic Theorist and later the General Problem Solver [47] led to the *logic modelling approach* to robotics. The most eminent example of this approach is 'Shakey', a robot constructed in 1966 that used logic reasoning for achieving its goals [56]. The robot had logic models of the current world state, of the actions it could take and their effects, and of the desired world state. It would then use logic reasoning to find the correct sequence of actions transforming the initial world state into the desired world state.

Figure 7.1 shows a simplistic example of such logic modeling. The robot's goal is not to be hungry (~hungry). To achieve the goal, it has a few actions at its disposal

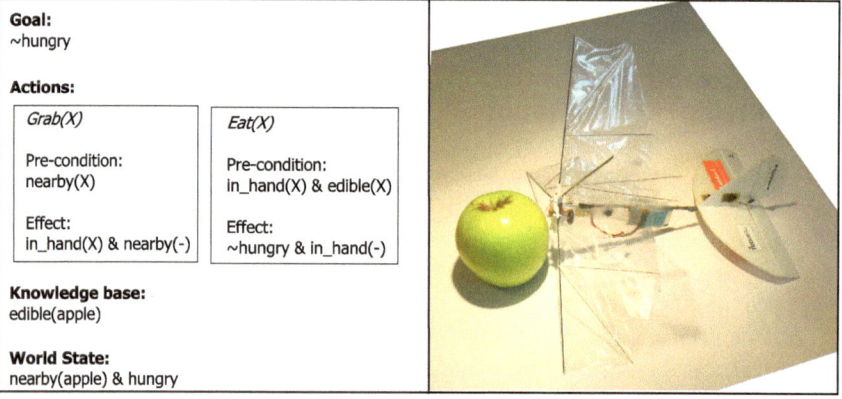

Fig. 7.1 Illustration of the logic modelling approach to robotics. See text for further details

(Grab(X) and Eat(X)). Each action has a pre-condition, which must be satisfied in order for the robot to be able to perform the action. The effect of the action is also represented in logic propositions. As an example, the action Eat(X) requires the presence of an edible object (edible(X)) in the hand of the robot (in_hand(X)). The presence of X in both pre-conditions implies that it is the same object. While some propositions depend on the world state, others are contained in the knowledge base. In this example, the robot knows that the object represented as 'apple' is edible. With its knowledge of the world and its actions, the robot can follow a systematic procedure to arrive at its goal. We let it to the reader to find out which actions the robot should take to still its hunger.

The logic modeling approach received criticism in the early 1980s. Sometimes the robots would spend a large amount of time reasoning about the next best action, due to a combinatorial explosion of possible actions and future world states. In addition, the execution of a plan was hampered, because the robot's models of the world and its actions did not correspond to reality. The logic models allowed no room for uncertainty on the world state.

One of the main scientists criticizing the logic modeling approach was Brooks, who argued for a *behavioral approach* to robotics [12,13]. The approach focused on the fact that robots, just like animals, need to react quickly to the environment. Logic reasoning was abandoned in exchange for multiple sensory-motor processes that were executed in parallel. In particular, Brooks introduced the subsumption architecture, in which there is a hierarchy of behavioral layers. All layers run in parallel, but higher layers can 'subsume' lower layers by suppressing and replacing their outputs. The parallel nature of control leads to the graceful decay of behavior in the case of failure. For instance, in [25] a subsumption architecture was made for the locomotion of a six-legged robot. The low-level behaviors were implemented on separate microcontrollers for all of the legs. A failure of one microcontroller then does not break down the robot's capability to move.

Figure 7.2 shows a small example of a subsumption architecture for a garbage collecting robot. The bottom layer lets the robot perform a reactive obstacle avoidance behavior. The middle layer takes care of navigation, so that the robot collects all garbage cans in its environment. The top layer is responsible for the actual collection and disposal of garbage cans. It clearly has to subsume the obstacle avoidance layer, since the robot will have to approach and touch the garbage cans in order to collect them.

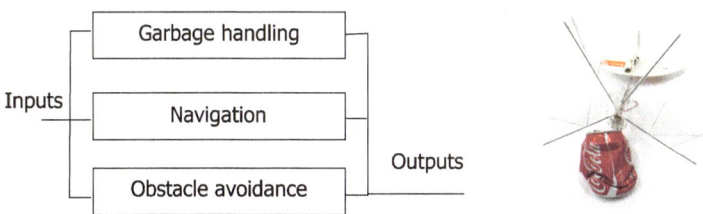

Fig. 7.2 An example of a small subsumption architecture for garbage handling

Fig. 7.3 An example of purposive vision, in which distances to obstacles in view are determined by means of their *y*-coordinate in the image. The picture was taken with a camera mounted on an RC-car

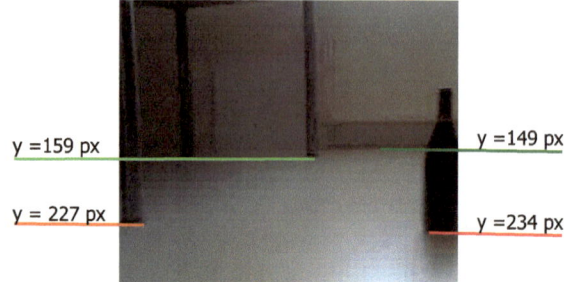

The behavioral approach built upon early work by Walter [67] and Braitenberg [11]. They showed that tasks such as phototaxis can be executed by robots with extremely simple controllers—without the need for representing the environment or the robot itself. The tasks of these robots were achieved by smart design of the link between sensory inputs and motor actions, or in other words *sensory-motor coordination*. Even for the important modality of vision, sensory-motor coordination can play an important role. For instance, in [2] it was shown that many hard vision problems such as shape from shading and structure from motion are much easier for an observer that can actively determine its viewpoints. This visual sensory motor-coordination is known as *active vision* or *animate vision* [2,5].

In related articles, the approach of *purposive vision* was forwarded [1,53]. According to this approach, vision is tightly linked to the observer's body, environment, and task. In particular, the robot should only extract the information from the environment that allows it to achieve its task. An example algorithm that stems from this approach is the algorithm that determines the distances to obstacles for ground robots (see Fig. 7.3). It does so by looking at which *y*-coordinate the obstacles touch the ground in the image. Closer obstacles touch the ground lower in the image. Of course, the robot is assumed to be situated on a flat ground plane with a fixed (or known) attitude.

Although the purposive vision approach has had quite some influence, it has a few limitations. First, the task-specificity makes the approach rather 'ad hoc' and hardly recognizable as an approach. Second, it is often not clear what information is necessary for a task, and how to extract this information only to the extent necessary for the task. Finally, due to the complexity of the closed loop nature of the sensory-motor apparatus, it turns out to be rather difficult to successfully design sensory-motor coordination by hand.

The approach of *evolutionary robotics* [48] draws inspiration from natural evolution to solve the above-mentioned problems. Evolutionary Algorithms (EAs) are used to optimize the robot's controller and sometimes body for its specific embodiment, task, and environment. Generally, a neural network is used as the controller, of which the weights and other parameters are optimized by the EA. This approach has shown for many different tasks that computationally efficient neural networks can implement successful sensory-motor coordination mechanisms. For example, in [49], a feedforward neural network was evolved to control a robot that had to approach small cylindrical objects and avoid big ones. When seen in a passive perspective, this task is very difficult, because many of the possible sensory inputs can occur for both types of objects. Indeed, collecting sensory inputs around both types of objects led to a large overlap of the two classes in the sensory space, and therefore to a rather bad classification performance. Good classification was only possible in a few locations around the objects. The evolved neural network controlled the robot in such a way that it would always end up at these favorable locations, allowing it to correctly approach or avoid them.

More recently, a *developmental* approach to robotics has been introduced (sometimes referred to as 'epigenetic robotics') [40,50]. This approach is very similar in philosophy to evolutionary robotics, but it has an emphasis on robotic development during the 'lifetime' of the robot (online learning). Examples of developmental robotics studies include the development of language [46], joint attention [36], and the discovery of the robot's own body [43] or visual experiences [8].

The example studies for evolutionary robotics and developmental robotics show a focus on demonstrating fundamental aspects of intelligence, not necessarily on engineering the best short-term solutions for robot applications.

While the behavior based and ensuing approaches to robotics can be seen as a counter reaction to the logic modeling approach, the current mainstream approach to robotics reconciles the description of the environment with the uncertainties of the world. This *probabilistic approach* to robotics explicitly models the uncertainty on the world state [14,63]. The most well-known method of this approach is Simultaneous Localization And Mapping (SLAM), in which the robot incrementally constructs a model of its environment, taking into account uncertainties in sensing and acting. Figure 7.4 shows a buggy carrying various sensors, including three different laser scanners, GPS, and inertial sensors (left), and a very accurate reconstruction of the environment (right) [10]. In SLAM, the robot typically maintains various hypotheses

Fig. 7.4 Illustration of an accurate reconstruction of an environment with SLAM [10]. *Left* Buggy with various sensors such as three laser scanners, GPS, IMU, and cameras. *Right* Reconstruction of a parking lot, with the *green line* indicating the trajectory of the buggy. Reprinted with kind permission from Springer science and business media

with associated probabilities, which is especially helpful if the robot is 'kidnapped' and moved to another part of the environment. Similar probabilistic algorithms were at the heart of robots that won the DARPA challenges for autonomous cars [64,65].

7.2 Challenges for Autonomous Flight

Most of the research mentioned in the previous section has been performed on ground robots, which can often carry quite some energy, sensors, and processing power onboard. The autonomous cars in the DARPA challenge are an extreme example: they can carry high-end laser scanners, multiple cameras, and multiple PCs on board for the sensing and computing. The autonomy of light-weight flying robots poses a larger challenge. In comparison with ground robots, Micro Air Vehicles (MAVs) can carry little energy and fewer, lower-quality sensors, while having far less processing power onboard. In addition, flying requires continuous fast reactions in order to avoid collisions.

We discern four levels of autonomous flight:

1. *Attitude control*: keeping the roll, pitch, and yaw angles within acceptable boundaries for sustained flight. This is often referred to as inner loop control.
2. *Height control*: keeping a desired height. We employ the definition here that *height* is the distance in the vertical direction to the terrain/objects underneath, while *altitude* is the distance to the Mean Sea Level (MSL), as can be estimated with barometric pressure. Keeping the altitude constant will not ensure the avoidance of (objects on) the ground.
3. *Collision avoidance*: In cluttered outdoor areas and in indoor areas this capability is essential, especially for platforms that are not very collision-resilient. Please note that collision avoidance is very much related to height control, since the latter can be interpreted as avoiding collisions in the vertical direction.
4. *Navigation*: being able to determine where to go in possibly unknown environments, in order to perform the robot's task.[1]

7.3 Approaches to Autonomous Flight

In open, outdoor areas, attitude control, altitude control and navigation are typically achieved by using both an Inertial Measurement Unit (IMU) and a Global Positioning System receiver (GPS) [6,51,66]. However, these sensors do not provide any information useful for height control and collision avoidance.

---

[1] Please note that this definition is very different from that in the control systems community, where navigation pertains to sensing the state of the system.

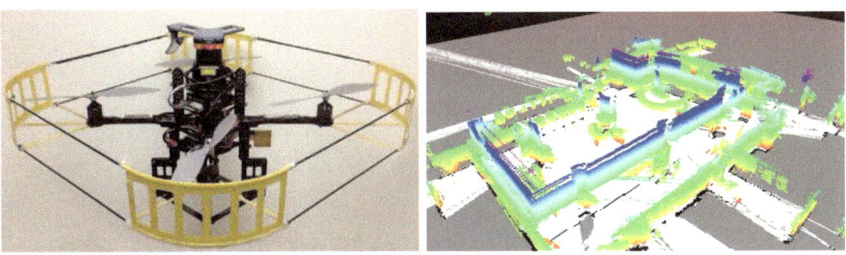

Fig. 7.5 Simultaneous Localization And Mapping on a quadrotor equipped with a laser scanner, from [22]. Reprinted with kind permission from Springer science and business media

There has been extensive research on collision avoidance. Notably, earlier research on larger UAVs with high-resolution laser scanners [58,61] provided promising results for navigation in cluttered environments. However, these UAVs weighed more than 75 kg and had to use most of their payload capability to lift the laser scanner. Currently, laser scanners are small enough for use on MAVs weighing around 1 kg by sacrificing both resolution and sensing directions. Scanners that measure distances to obstacles in a 2D plane through the MAV are part of the most successful systems for indoor flight [3,30]. Figure 7.5 shows results from an investigation of SLAM for MAVs [22]. The left part of the figure shows the quadrotor equipped with a minia-turized laser scanner. The right part shows the 3D-map that has been constructed with SLAM on the basis of many different readings. The map is a point cloud that is colored according to the height.

In a similar vein, the Microsoft Kinect has been adapted for use on MAVs [23,34,59]. The main disadvantage of sensors such as the laser scanner or Kinect is that they are *active* sensors, i.e., they measure distances by actively sending out signals. This implies that they consume quite some energy, reducing the MAV's flight duration considerably.

There is a large body of research on autonomous flight by using cameras. Cameras are *passive* sensors and as such consume much less energy than active sensors such as laser scanners. In addition, they provide information about a large part of the environment at once, including obstacles at large distances. Moreover, cameras can be made very light weight, up to the order of milligrams. The challenge of using cameras is to design computer vision algorithms that extract useful information from the images in a computationally efficient manner.

We identify two main approaches to vision-based autonomous flight. The first one follows the probabilistic approach to robotics. Typically, methods from projective geometry [31] are used to determine the motion of the camera over time and the 3D-structure of the environment. This can be done by matching features from images over time. In *visual odometry* the main interest is in the motion of the camera over time, while in visual SLAM the goal is to have a persistent 3D-map of the environment that can eliminate long-term drift cf. [18,27,28,68]. In SLAM, the construction of an accurate 3D-model of the world is part of an attempt to solve collision avoidance

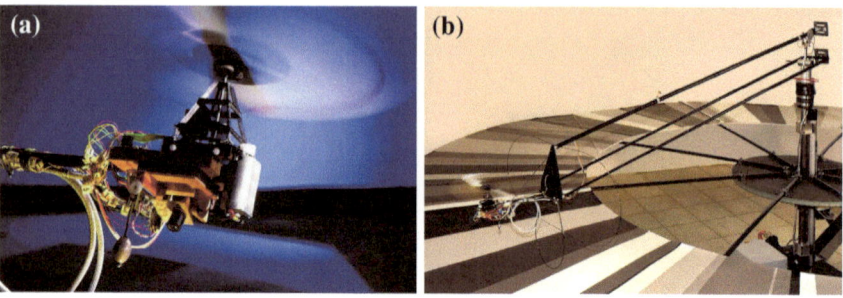

Fig. 7.6 Tethered UAV with tiny Elementary Motion Detectors that is able to navigate and land on a moving platform—publicly available image from [54]

and navigation in the same time. This leads to a rather complex problem setting. Despite various recent speed-ups of the involved algorithms [60,62], this makes the algorithms computationally rather expensive.

The second method follows the behavioral approach to robotics. It draws inspiration from biological systems such as honeybees, which with their limited sensory apparatus and brain (only ∼960,000 neurons [45]) can perform daunting navigation tasks. Studies implementing this method typically abandon a state estimate altogether and focus on designing the right responses to incoming visual inputs [9,38]. As a consequence, the designed controllers for autonomous flight are computationally extremely efficient. This allows autonomous flight of very light-weight platforms, such as the 10 g indoor flyer from [69], which could avoid the walls in a strongly textured environment. The sensor readings generally consist of optical flow complemented with the rotation rates [17,55], since these have been shown to play an important role in insect flight [16]. Figure 7.6 shows an experimental setup in which artificial Elementary Motion Detectors (EMDs) are used to navigate and land on a moving platform [54]. However, even for advanced optical flow sensors [24,26,57] the optics and the optical flow algorithms of flying robots have a lower performance than their natural counterparts. As a consequence, the success of optical flow algorithms relies on the presence of sufficient texture in the environment. This poses limits on the environments in which the robots are able to fly.

7.3.1 Flapping Wing MAVs

Flapping wing MAVs are on the low-weight end of the spectrum of different MAV types. The low weight (e.g., 16 g for the DelFly II) implies that the power, sensor, and processing requirements become much more stringent than for a typical MAV of around 1 kg. Since the design of flapping wing MAVs is still an active area of research (see Chaps. 2–4), achieving enough lift for carrying a small camera on board is already a considerable challenge.

Consequently, studies on the autonomy of flapping wing MAVs typically focus on the use of external cameras, mostly for attitude and height control [15,33,39],

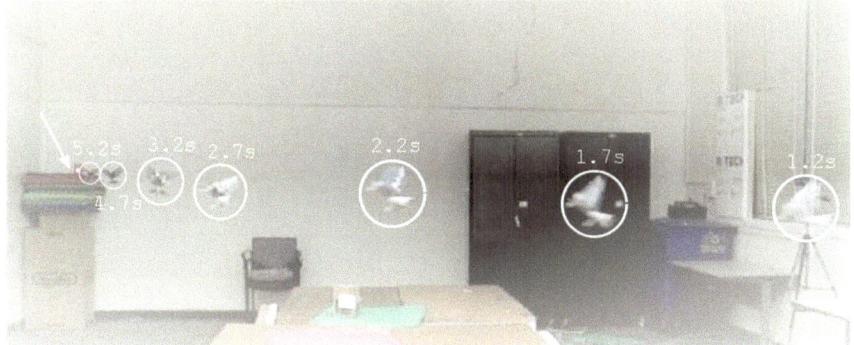

Fig. 7.7 A flapping wing MAV equipped with an onboard Wii-mote camera flies towards an infrared light, from [4]—reprinted with permission

but more recently also for the more complex task of passing through a window [35]. Obviously, the need for an external camera system prevents autonomous flight in unknown environments.

Attitude control can be performed with small onboard sensors. In the Nano Hummingbird, onboard gyro feedback was used to facilitate manual attitude control [37]. The tiny Robobee first attained free flight with the help of external cameras [41], but now also is equipped with onboard sensors such as accelerometers and gyros [29], or magnetometers [32] for attitude control.

The larger flapping wing MAV in [7] had both a camera and a processor on board, with which optical flow was calculated. In that study, the weight of the onboard sensors and processor was too high for the MAV to ascend, and only descents were considered. In later work of the same group the generated lift and the onboard electronics have been considerably improved. In [4] for the first time on a flapping wing MAV a camera with onboard processing was used. The flapping wing was equipped with a camera from the Wii remote [52] that automatically tracks an infrared light source. The MAV used it to perform the task of target finding (Fig. 7.7).

None of the above-mentioned studies applies onboard sensors to collision avoidance, which is a basic and essential capability for autonomous flight in unknown environments.

7.3.2 Approach to Autonomous Flight of the DelFly

The final goal of the DelFly project is to achieve autonomous flight of light-weight flapping wing MAVs in real-world, unknown environments. Of course, on such light-weight platforms, computational efficiency is of the utmost importance. Hence, it is useful to draw inspiration from natural systems. However, there are also important hardware differences between the DelFly and small flying insects. These differences affect the possibilities for sensing and acting, and have required us to go beyond mere copying of strategies found in flying insects.

Fig. 7.8 The DelFly II autonomously avoiding obstacles by determining whether it should turn left or right. *Left* External picture. *Right* Greyscale onboard image with determined optical flow vectors (*green lines*). From the experiments in [20]

Concerning the autonomy levels described in Sect. 7.2, attitude control has never been an issue. The DelFly's tail stabilizes the attitude in a slow forward flight regime. Moreover, navigation only becomes relevant after we can ensure safe short-term operation, implying that height control and collision avoidance have to be solved first.[2] These tasks are very difficult for a light-weight flapping wing MAV that generally moves at a speed of 0.5–1.0 m/s in narrow indoor spaces, and we do not yet consider them fully solved.

All DelFly models carried a camera, which has been the primary sensor for autonomous flight. Our main focus has been on how to extract as efficiently as possible visual information relevant to the DelFly's task. Hence, we have adhered to a purposive vision approach to autonomous flight. Various (relatively simple) visual cues have been studied, ranging from optical flow based time-to-contact estimates for collision avoidance [20] (see Fig. 7.8) to height classification based on visual appearances features [19].

In order to autonomously explore unknown environments without range constraints, the flapping wing MAV will have to capture and process images on board. However, onboard processing has been beyond our reach for a long time. Instead, almost all experiments with the DelFly have been performed with the onboard images being transmitted to a laptop computer for off-board image processing. This setup influences the research in two important ways. First, offboard processors are computationally more powerful than possible onboard processors, because they are not subject to the same weight and energy requirements. One has to take care that offboard implemented vision algorithms are computationally efficient enough, so that they have a chance of being implemented on a possible onboard processor. Second, the use of offboard computation also introduces problems. For instance, analog image transmission introduces delays and various types of image noise that are not present in an onboard processing scheme. In 2013, we have performed the first experiments

---

[2]With *navigation* we mean that the robot can decide where to go next in its environment in order to achieve its goals.

with not only onboard cameras but also onboard vision processing on a limited embedded processor.

As was mentioned in Sect. 7.1, a major problem with purposive vision is to determine what information is necessary for a task, and how to extract this information only to the extent necessary for the task. In this respect, we have made some interesting progress during the DelFly project. Most of the time the DelFly would determine its actions based on only a few variables, such as its estimated height or distance to an obstacle. These estimates are typically the result of processing different local parts of the image. For instance, in [19] the height was estimated on the basis of a number s of small image samples. Changing this number of samples allowed us to explore the trade-off between accuracy and computational efficiency. We have often found this tradeoff to be very benign; large speedups are possible at little cost in accuracy [21]. This rather generic method of *sub-sampling* allows the designer to optimally tune the extent to which the information has to be extracted for the task. It has been one of the key insights that allowed us to immediately port our algorithms to the onboard processor when it became available.

The extracted visual cues typically resulted in a very limited number of variables to be used by the controller, for which we have mainly employed Finite-State-Machines (FSMs, cf. [56]). FSM-controllers with a small number of states and transitions are extremely efficient in terms of computation and memory, while they allow the control to go beyond simple 'reflexes' that always map the same sensory inputs to the same actions. Although we have mainly used FSMs for the tasks of height control and obstacle avoidance, it is possible to modularize and extend FSM-controllers to achieve more complex tasks. In the end, we hope to reach the level of autonomous navigation by linking together relatively simple and computationally efficient modules in a smart manner.

The remainder of this part of the book is outlined as follows. Rather than providing detailed information on the autonomy experiments in chronological order, we will focus on some of the most significant contributions. We will first discuss monocular vision and then move to binocular vision. In Chap. 8 we explain how the DelFly uses a single camera to detect obstacles by complementing optic flow with a visual appearance cue. Subsequently, in Chap. 9, an investigation is performed to achieve a turning logic that is suited for unknown environments with little texture. Finally, we present in Chap. 10 how the switch to onboard stereo vision now allows autonomous flights in unknown environments for as long as the DelFly's battery lasts.

References

1. J. Aloimonos, Purposive and qualitative active vision. In: *10th International Conference on Pattern Recognition*, vol. 1, Atlantic City, NJ, pp. 346–360 (1990)
2. J. Aloimonos, I. Weiss, A. Bandyopadhyay, Active vision. Int. J. Comput. Vis. **1**(4), 333–356 (1988)

3. A. Bachrach, R. He, N. Roy, Autonomous flight in unstructured and unknown indoor environments, in *EMAV, The Netherlands* (2009)
4. S.S. Baek, Autonomous ornithopter flight with sensor-based behavior. Technical Report UCB/EECS-2011-65 (2011)
5. D.H. Ballard, Animate vision. Artif. Intell. **48**(1), 57–86 (1991)
6. R.W. Beard, D. Kingston, M. Quigley, D. Snyder, R.S. Christiansen, W. Johnson, Autonomous vehicle technologies for small fixed-wing UAVs. J. Aerosp. Comput. Inf. Commun. **2**(1), 92–108 (2005)
7. F.G. Bermudez, R. Fearing, Optical flow on a flapping wing robot, in *IROS 2009*, pp. 5027–5032 (2009)
8. O. Berthold, V.V. Hafner, Unsupervised learning of sensory primitives from optical flow fields, in *Simulation of Adaptive Behavior (SAB 2014), LNAI 8575*, pp. 188–197 (2014)
9. A. Beyeler, J-C. Zufferey, D. Floreano, 3d vision-based navigation for indoor microflyers, in *ICRA 2007, Italy*, pp. 1336–1341 (2007)
10. José-Luis Blanco, Francisco-Angel Moreno, Javier González, A collection of outdoor robotic datasets with centimeter-accuracy ground truth. Auton. Robots **27**(4), 327–351 (2009)
11. V. Braitenberg, *Vehicles: Experiments in Synthetic Psychology* (MIT Press, Cambridge, 1984)
12. R.A. Brooks, Elephants don't play chess. Robot. Auton. Syst. **6**(1–2), 3–15 (1990)
13. R.A. Brooks, Intelligence without representation. Artif. Intell. **47**(1–3), 139159 (1991)
14. W. Burgard, D. Fox, S. Thrun, Active mobile robot localization, in *14th International Joint Conference on Artificial Intelligence (IJCAI 1997)*, San Mateo, CA, Morgan Kaufmann (1997)
15. C.-L. Chen, F.-Y. Hsiao, Attitude acquisition using stereo-vision methodology, in *IASTED Conference* (2009)
16. T.S. Collett, Insect vision: controlling actions through optic flow. Curr. Biol. **12**, 615–617 (2002)
17. J. Conroy, G. Gremillion, B. Ranganathan, J.S. Humbert, Implementation of wide-field integration of optic flow for autonomous quadrotor navigation. Auton. Robots **27**(3), 189–198 (2009)
18. A.J. Davison, D.W. Murray, Simultaneous localisation and map-building using active vision, in *IEEE PAMI* (2002)
19. G.C.H.E. de Croon, C. de Wagter, B.D.W. Remes, R. Ruijsink, Random sampling for indoor flight, in *International Micro Air Vehicle conference* (Braunschweig, Germany, 2010), p. 2010
20. G.C.H.E. de Croon, M.A. Groen, C. De Wagter, B.D.W. Remes, R. Ruijsink, B.W. van Oudheusden. Design, aerodynamics, and autonomy of the delfly. Bioinspir. Biomimet. **7**(2) (2012)
21. G.C.H.E. de Croon, C. De Wagter, B.D.W. Remes, R. Ruijsink, Sub-sampling: real-time vision for micro air vehicles. Robot. Auton. Syst. **60**(2), 167–181
22. I. Dryanovski, R.G. Valenti, J. Xiao, An open-source navigation system for micro aerial vehicles. Auton. Robots **34**(3), 177–188 (2013)
23. F. Endres, J. Hess, J. Sturm, D. Cremers, W. Burgard, 3d mapping with an RGB-D camera, in *IEEE Transactions on Robotics (T-RO)* (2013)
24. F. Expert, S. Viollet, F. Ruffier, Outdoor field performances of insect-based visual motion sensors. J. Field Robotics **28**(4), 529–541 (2011)
25. C. Ferrell, Robust agent control of an autonomous robot with many sensors and actuators. Master's thesis, Electrical Engineering and Computer Science, Massachusetts Institute of Technology (1993)
26. D. Floreano, R. Pericet-Camara, S. Viollet, F. Ruffier, A. Brückner, R. Leitel, W. Buss, M. Menouni, F. Expert, R. Juston, M.K. Dobrzynski, G. LEplattenier, F. Recktenwald, H.A. Mallot, N. Franceschini, Miniature curved artificial compound eyes. Proc. Natl. Acad. Sci. USA, PNAS **110**(23), 9267–9272 (2013)
27. C. Forster, M. Pizzoli, D. Scaramuzza, Svo: fast semi-direct monocular visual odometry, in *IEEE International Conference on Robotics and Automation (ICRA)* (2014)

28. F. Fraundorfer, D. Scaramuzza, Visual odometry: Part ii—matching, robustness, and applications. IEEE Robot. Automat. Mag. **19**(2) (2012)

29. S.B. Fuller, E.F. Helbling, P. Chirarattananon, R.J. Wood, Using a mems gyroscope to stabilize the attitude of a fly-sized hovering robot, in *IMAV 2014: International Micro Air Vehicle Conference and Competition* (2014)

30. S. Grzonka, G. Grisetti, W. Burgard, Towards a navigation system for autonomous indoor flying. In *(ICRA 2009), Kobe, Japan* (2009)

31. R.I. Hartley, A. Zisserman, *Multiple View Geometry in Computer Vision*, 2nd edn. (Cambridge University Press, Cambridge, 2004)

32. E.F. Helbling, S.B. Fuller, R.J. Wood, Pitch and yaw control of a robotic insect using an onboard magnetometer. in *IEEE International Conference on Robotics and Automation (ICRA)*, pp. 5516–5522 (2014)

33. F.Y. Hsiao, H.K. Hsu, C.L. Chen, L.J. Yang, J.F. Shen, Using stereo vision to acquire the flight information of flapping-wing mavs. J. Appl. Sci. Eng. **15**(3), 213–226 (2012)

34. A.S. Huang, A. Bachrach, P. Henry, M. Krainin, D. Maturana, D. Fox, N. Roy, Visual odometry and mapping for autonomous flight using an RGB-D camera. in *International Symposium on Robotics Research (ISRR)* (2011)

35. R.C. Julian, C.J. Rose, H. Hu, R.S. Fearing, Cooperative control and modeling for narrow passage traversal with an ornithopter mav and lightweight ground station, in *12th International Conference on Autonomous Agents and Multiagent Systems (AAMAS 2013)* (2013)

36. F. Kaplan, V.V. Hafner, The challenges of joint attention, in *Proceedings of the Fourth International Workshop on Epigenetic Robotics* (2004)

37. M. Keennon, K. Klingebiel, H. Won, and A. Andriukov. Development of the nano hummingbird: a tailless flapping wing micro air vehicle, in *50th AIAA Aerospace Science Meeting*, pp. 6–12 (2012)

38. S. Leven, J.-C. Zufferey, D. Floreano, A minimalist control strategy for small UAVs, in *(IROS 2009)*, pp. 2873–2878 (2009)

39. S.H. Lin, F.Y. Hsiao, C.L. Chen, J.F. Shen, Altitude control of flapping-wing mav using vision-based navigation, in *IEEE International Conference on Robotics and Automation, 2009. ICRA '09.*, pp. 3644–3650 (2009)

40. M. Lungarella, G. Mettay, R. Pfeifer, G. Sandini, Developmental robotics: a survey. Connect. Sci. **15**(4), 151–190 (2003)

41. K.Y. Ma, P. Chirarattananon, S.B. Fuller, R.J. Wood, Controlled flight of a biologically inspired, insect-scale robot. Science **340**(6132), 603–607 (2013)

42. A.P. Mahon, The history of hut eight 1939–1945. UK National Archives Reference HW 25/2. http://www.ellsbury.com/hut8/hut8-000.htm (retrieved 1-12-2014) (1945)

43. G. Martius, L. Jahn, H. Hauser, V.V. Hafner, Self-exploration of the stumpy robot with predictive information maximization, in *From Animals to Animats 13*, vol. 8575, pp. 32–42 (2014)

44. P. McCorduck, *Machines Who Think*, 2nd edn. (A. K. Peters Ltd, Natick, 2004)

45. R. Menzel, M. Giurfa, Cognitive architecture of a mini-brain: the honeybee. Trends Cogn. Sci. **5**(2), 62–71 (2001)

46. A.F. Morse, P. Baxter, T. Belpaeme, L.B. Smith, A. Cangelosi, The power of words, in *IEEE International Conference on Development and Learning and Epigenetic Robotics (ICDL-EpiRob)* (2011)

47. A. Newell, J.C. Shaw, H.A. Simon, Report on a general problem-solving program, in *Proceedings of the International Conference on Information Processing*, pp. 256–264 (1959)

48. S. Nolfi, D. Floreano, *Evolutionary Robotics: The Biology, Intelligence, and Technology of Self-Organizing Machines* (MIT Press/Bradford Books, Cambridge, 2000)

49. S. Nolfi, D. Marocco, Evolving robots able to visually discriminate between objects with different size. Int. J. Robot. Automat. **17**(4), 163–170 (2002)

50. P.-Y. Oudeyer, On the impact of robotics in behavioral and cognitive sciences: from insect navigation to human cognitive development. IEEE Trans. Auton. Mental Dev. **2**(1) (2010)

51. B.D.W. Remes, P. Esden-Tempski, F. van Tienen, E.J.J. Smeur, C. De Wagter, G.C.H.E. de Croon, Lisa-s 2.8g autopilot for gps-based flight of mavs, in *International Micro Air Vehicle conference and competitions (IMAV 2014)* (2014)

52. Nintendo Wii Remote. https://en.wikipedia.org/wiki/Wii_Remote

53. E. Rivlin, Y. Aloimonos, Purposive active vision: combining perception and action, in *International AI Symposium* (1992)

54. F. Ruffier, N. Franceschini, Optic flow regulation in unsteady environments: a tethered mav achieves terrain following and targeted landing over a moving platform. J. Intell. Robot. Syst. (2014)

55. F. Ruffier, N.H. Franceschini, Aerial robot piloted in steep relief by optic flow sensors, in *(IROS 2008)*, pp. 1266–1273 (2008)

56. S.J. Russell, P. Norvig, *Artificial Intelligence: A Modern Approach*, 3rd edn. (Prentice Hall, Upper Saddle River, 2010)

57. G. Sabiron, P. Chavent, T. Raharijaona, P. Fabiani, F. Ruffier, Low-speed optic-flow sensor onboard an unmanned helicopter flying outside over fields, in *IEEE International Conference on Robotics and Automation* (2013)

58. S. Scherer, S. Singh, L. Chamberlain, S. Saripalli, Flying fast and low among obstacles, in *Proceedings International Conference on Robotics and Automation* (2007)

59. S. Shen, N. Michael, V. Kumar, A stochastic differential equation-based exploration algorithm for autonomous indoor 3d exploration with a micro-aerial vehicle. Int. J. Robot. Res. **31**(12), 1431–1444 (2012)

60. S. Shen, Y. Mulgaonkar, N. Michael, V. Kumar, Vision-based state estimation and trajectory control towards high-speed flight with a quadrotor, in *Robotics Science and Systems IX (RSS 2013)* (2013)

61. D.H. Shim, H. Chung, H.J. Kim, S. Sastry, Autonomous exploration in unknown urban environments for unmanned aerial vehicles, in *AIAA GNC Conference, San Francisco* (2005)

62. F. Steinbrucker, J. Sturm, D. Cremers, Volumetric 3d mapping in real-time on a CPU, in *2014 IEEE International Conference on Robotics and Automation (ICRA)*, IEEE, pp. 2021–2028 (2014)

63. S. Thrun, Probabilistic algorithms in robotics. AI Mag. **21**(4), 93–109 (2000)

64. S. Thrun, M. Montemerlo, H. Dahlkamp, D. Stavens, A. Aron, J. Diebel, P. Fong, J. Gale, M. Halpenny, G. Hoffmann, K. Lau, C. Oakley, M. Palatucci, V. Pratt, P. Stang, S. Strohband, C. Dupont, L.-E. Jendrossek, C. Koelen, C. Markey, C. Rummel, J. van Niekerk, E. Jensen, P. Alessandrini, G. Bradski, B. Davies, S. Ettinger, A. Kaehler, A. Nefian, P. Mahoney, Stanley: the robot that won the darpa grand challenge. J. Field Robot. **23**(9), 661–692 (2006)

65. C. Urmson, J. Anhalt, J. Andrew (Drew) Bagnell, C.R. Baker, R.E. Bittner, J.M. Dolan, D. Duggins, D. Ferguson , T. Galatali, H. Geyer, M. Gittleman, S. Harbaugh, M. Hebert, T. Howard, A. Kelly, D. Kohanbash, M. Likhachev, N. Miller, K. Peterson, R. Rajkumar, P. Rybski, B. Salesky, S. Scherer, Y.-W. Seo, R. Simmons, S. Singh, J.M. Snider, A. (Tony) Stentz, W. (Red)L. Whittaker, J. Ziglar, Tartan racing: a multi-modal approach to the darpa urban challenge. Technical Report CMU-RI-TR-, Robotics Institute. http://archive.darpa.mil/grandchallenge/ (April 2007)

66. K.P. Valavanis, *Advances in Unmanned Aerial Vehicles* (Springer, New York, 2007)

67. W.G. Walter, A machine that learns. Sci. Am. **185**, 60–63 (1951)

68. B. Williams, I.D. Reid, On combining visual slam and visual odometry, in *IEEE International Conference on Robotics and Automation* (2010)

69. J.-C. Zufferey, *Bio-inspired Flying Robots: Experimental Synthesis of Autonomous Indoor Flyers* (EPFL/CRC Press, Lausanne, 2008)

Monocular Obstacle Detection

8

Abstract

This chapter deals with monocular obstacle detection. The visual cue of optic flow is used to determine the time-to-impact to obstacles in the environment. Since the flapping wing motion hampers the determination of optic flow, a complementary, "appearance variation cue" is studied. Combining these visual cues significantly improves detection results.

8.1 Introduction

In this chapter, we will focus on the algorithms used by the DelFly to detect obstacles. As mentioned in Chap. 7, the primary visual cue used for obstacle detection in the literature is optic flow. When approaching an obstacle, the optic flow increasingly expands. With a few simplifying assumptions, the rate of expansion can be used to estimate the time-to-impact to the objects in view (e.g., [16,22]).

While optic flow is commonly used in efforts for reaching autonomous flight, image appearance has been largely neglected. Image appearance features could be useful for autonomous indoor flight, since they can capture information complementary to optic flow. For example, the absence of texture (a fail-case for optic flow) can be successfully detected by extracting appearance features. The little interest in appearance features is mainly due to the fact that their extraction is computationally expensive (Fig. 8.1).

In order to improve the obstacle detection capabilities of the DelFly, a novel appearance cue for obstacle detection is investigated [7]. The cue captures the variation in texture and/or color in a single image, and is based on the assumption that there is less such variation when the camera is close to an obstacle. Henceforth, the cue is referred to as the *appearance variation cue*. A naive implementation of the appearance variation cue would be computationally expensive. *Sub-sampling* is employed for significantly reducing the computational effort of extracting image appearance

G.C.H.E. de Croon et al., *The DelFly*, DOI 10.1007/978-94-017-9208-0_8

Fig. 8.1   The *DelFly Micro* is a 3.07-gram ornithopter with a wing span of 10 cm. Autonomous flight will require fast image processing with little computational power available

features. The higher computational efficiency comes at the cost of a lower accuracy. The key variable that determines the trade-off between efficiency and accuracy is the number of samples extracted from the image.

The remainder of the chapter is organized as follows. In Sect. 8.2, the novel appearance cue is investigated. Subsequently, in Sect. 8.3 the strategy of sub-sampling is analyzed in terms of its computational efficiency. The resulting accuracy of the method is analyzed in Sect. 8.3.3. In Sect. 8.4 it is shown that the sampled novel cue is complementary to the time-to-impact determined by optic flow on a classification task; the classification performance increases when the two visual cues are used together. In Sect. 8.6 closed-loop experiments are discussed, in which the flapping wing MAV DelFly II autonomously avoids obstacles using the appearance cue and optic flow. Finally, we discuss the results in Sect. 8.7 and draw conclusions in Sect. 8.8.

8.2   Appearance Variation as a Cue for Obstacle Proximity

Humans use various cues to determine distances to objects in their environment. Monocular visual cues include [3, 14, 21]:

- Apparent motion: when the observer moves relative to the environment, different objects in the environment appear to move in different directions and with different speeds. Notably, in case of a translation of the observer, closer objects have a larger apparent motion than those further away.
- Perspective: parallel lines in the environment can intersect in the 2D-projection of the observer's 'image', providing cues about the relative distance between two parts of an object or the observer's placement in the environment.
- Object size: the sizes of familiar objects in the image provide information on their distance to the observer. The change of an object's size over time is an apparent motion cue for its time-to-impact with the observer.
- Aerial perspective: in a large environment, far-away objects have lower luminance contrast and color saturation. This cue is also termed 'distance fog'.

- Accomodation of the lens/blur: when focusing on objects at different distances, the lens of the eye accomodates to the distance, bringing the object in focus. This oculomotor cue together with the visual blur of objects at distances that are out of focus provides information on their distance to the observer.
- Occlusion: depending on the point of view of the observer, closer objects can occlude, i.e. hide, parts of other objects that are further away from the observer.
- Texture gradient: the texture of an object or the ground is finer at larger distances from the observer.
- Shadows: if the position of the light source relative to the observer is known, a shadow casted by one object on another can provide information on their relative distances to the observer.
- Image height cues: below the horizon a higher position at which an object touches the ground in the image signifies a larger distance to the observer.

Roboticists have mostly focused on the first cue for obstacle detection. Current optic flow algorithms are quite robust and computationally efficient in determining the motions of different visual elements over time. The subsequent processing of the resulting image motion vectors can range from rather straightforward (and computationally efficient) to more complex (and computationally more demanding). Reasons for the less frequent use of the other cues are the increased difficulty and computational effort for extracting them.

Here, a novel cue for estimating obstacle proximity is investigated for use in robotics. It is termed the *appearance variation cue*. When an observer approaches an object, there are two concurrent effects:

1. The image size of the object increases in the image, while other objects go out of view.
2. The detailed texture of the object in view becomes more and more visible.

The *main assumption underlying the novel cue* is that the variation in appearance of many different objects in view is larger than that of the detailed texture of one object alone. In other words, it is assumed that in general the first effect decreases the appearance variation more than the second effect increases it.

The appearance variation cue pertains to a single static view of a scene. The cue depends on the distance and on the textures and colors of the objects in view. *We expect the cue to be complementary to optic flow*, since (i) it is fit for dealing with the absence of texture, a fail-case for optic flow, (ii) it directly depends on the distance and does not require motion of the observer, and (iii) it does not require accurate, sub-pixel measurements in the image, which can make it more robust to noise and other degradations of the image. The verification of the expected complementarity with optic flow is performed in Sect. 8.4.

In this section, we verify whether the appearance variation indeed decreases towards impact. Section 8.2.1 explains the algorithm that measures the appearance variation. Then, Sect. 8.2.2 explains the experimental setup for testing the algorithm and Sect. 8.2.3 contains the results.

8.2.1 Measuring Appearance Variation

For measuring the appearance variation, the term 'appearance' is interpreted as textures and/or colors. The approach to estimating the 'variation' of these properties is to first estimate the probability distribution of the occurrence of different textures and/or colors in the image. Subsequently, the Shannon entropy [19] of the estimated probability distribution is calculated. Given a discrete probability distribution p with probabilities for n different 'events' p_i, the Shannon entropy $H(p)$ expressed in bits is:

$$H(p) = -\sum_{i=1}^{n} p_i \log_2(p_i), \qquad (8.1)$$

where $p_i \log_2(p_i) = 0$ for $p_i = 0$. A high entropy corresponds to a high variation in appearance, while a low entropy corresponds to a low variation.

There are many computer vision algorithms that can evaluate the distribution of texture and/or color in an image. In this section, two such methods are employed: the texton method [24] and a color distribution.

8.2.1.1 Texton Method for Estimating the Texture/Color Distribution

The texton method [24] is a computationally efficient method for representing the texture and/or color in an image. The method constructs a distribution on the basis of small local image samples. It was shown to be superior to computationally intensive filtering methods (e.g., Gabor filters) on a texture classification task. In addition, the texton method is amenable to the sub-sampling approach explained in Sect. 8.3. Below, we describe the implementation of the texton method in our experiments.

The texton method starts with the formation of a 'dictionary' of n textons.[1] To this end, small image samples of size $w \times h$ pixels are extracted from a set of images. This image set can be selected offline, before operation, or can be automatically gathered during operation. The extracted samples are clustered by means of a Kohonen network [15]. There are other - more advanced - clustering techniques [23], but the Kohonen network has the advantage of learning the clusters in an iterative fashion. It finds clusters quickly, and can be stopped as soon as the clusters seem to cover the different samples sufficiently.

After learning the dictionary, the texton method evaluates texture by estimating the probability distribution of different textons in the image. s image samples are extracted from the image to build a histogram g. For each sample, the closest texton i in the dictionary is determined (Euclidian distance), and the corresponding bin in the histogram g_i is incremented. Normalizing g results in a maximum likelihood estimate $\hat{p}$ of texton occurrence in the image, with $\hat{p}_i = g_i/s$. This estimate is inserted into (8.1) to determine the texture variation.

Two aspects of the method are worth mentioning. First, typically all possible local samples are extracted from the image, making $\hat{p}$ equal to p. Second, the texton

---

[1] All parameter settings will be mentioned in Sect. 8.2.2.

method captures only texture when it is applied to grayscale images, but it captures both texture and color when it is applied to color images. In the latter case, the textons actually have the dimension $w \times h \times 3$. In this section, both the grayscale and color version are studied.

8.2.1.2   Color Distribution

For determining the color distribution, the image is first transformed to the *Hue Saturation Value* (HSV) image space. Then, all pixels of the image are evaluated in order to construct a histogram of the different colors. A pixel is only evaluated if its S and V values exceed the corresponding thresholds ϑ_S and ϑ_V. If it exceeds these thresholds, the pixel is binned in one of n bins, according to the H value. Since the domain of H is $[0, 1]$ each bin has size $1/n$. If a pixel value does not exceed the thresholds, it is binned in an extra 'colorless' bin. The total number of bins is therefore $n + 1$. The resulting histogram is normalized to obtain a maximum likelihood estimate of the color probability distribution. Again, since all pixels are used, $\hat{p} = p$.

8.2.2   Experimental Setup

Three video sets are used in the experiments. First, we determine the appearance variation for a set of obstacle approach sequences made with an analog wireless camera. Second, a larger set of approach sequences is studied, which are captured with the camera of a mobile phone. Third, to study an even larger number of approach sequences, we also simulate approach sequences by zooming in on digital photographs. Of course, the disadvantages of simulated approaches are the absence of (a) threedimensional visual effects and (b) realistic noise conditions.

For the filmed approach sequences, a set of 10 videos was made by holding the analog wireless camera by hand and a set of 65 videos was made in the same manner but then with the mobile phone. Each approach starts from a distance of 3 m from a wall or other large obstacle, which is approached with an approximately constant pace. The different sequences range in length from 48 to 88 image frames. All images are resized to 160×120 pixels. The top row of Fig. 8.2 shows shots from three of the sequences.

For the simulated approaches, a set of 62 photographs was taken in inside environments at a distance of 3 m from a wall or other large obstacle. All photographs have a resolution of 3072×2304 pixels. The simulated camera has a resolution of 160×120 pixels. At the begin of the approach sequence the camera captures an area of 2000×1500 pixels, which is resized to the camera resolution with bicubic sampling. At the end of the approach, the captured area equals the camera resolution. Digitally zooming in further would always result in a lower appearance variation, since no new details can come into sight. We generate 310 image sequences, all comprising 90 frames during which the width and height of the captured image area

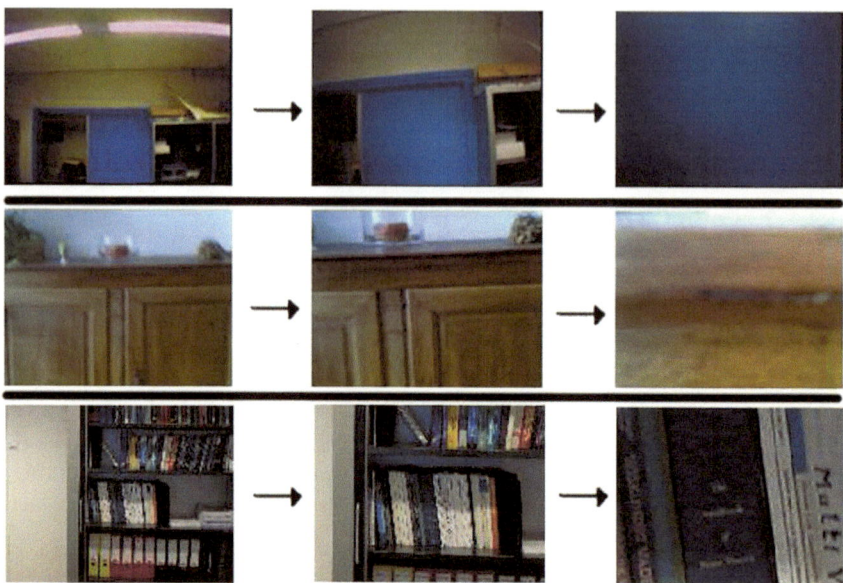

Fig. 8.2 Example images from a filmed approach sequence captured with a wireless camera (*top*), one with a mobile phone (*middle*) and a simulated approach sequence (*bottom*)

decreased linearly. The bottom row of Fig. 8.2 shows three images from one of the sequences.

Three algorithms were applied to the approach sequences: (1) the Texton Method with a Gray-scale dictionary (*TMG*), (2) the Texton Method with a Color dictionary (*TMG*), and (3) the Color Distribution Method (*CDM*). The settings of TMG are as follows. The size of the image patches is $w \times h = 5 \times 5$, and it extracts all possible samples in the gray-scale image, $s = 155 \times 115 = 17825$. The dictionary has $n = 30$ textons and it is learned on a separate set of images similar to those used for the experiment. TMC had the same settings as TMG but with texton sizes $5 \times 5 \times 3$. CDM had its thresholds set to $\vartheta_S = 0.1$ and $\vartheta_V = 0.1$, and used $n = 20$. This number of bins gave better results on preliminary experiments than $n = 30$.

8.2.3   Results

Figure 8.3 shows the entropy values of TMG over time (thin gray lines) for all filmed video sequences. The x-axis represents the time, while the y-axis shows the entropy. It also shows linear fits (thick blue lines), determined with a least squares optimization method.

As can be seen in Fig. 8.3, the slopes of the linear fits of the entropy over time are all negative. In order to get an idea of how often the entropy decreases towards impact, the proportions of negative slopes are determined for the other methods and video

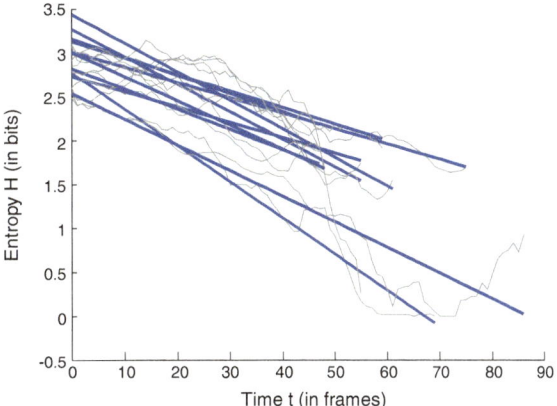

Fig. 8.3 The entropy of TMG over time (*light gray lines*) and corresponding linear fits (*thick blue lines*). The entropy decreases towards impact

Table 8.1 Proportion of decreasing appearance variation for the methods TMG, TMC, and CDM

	TMG (%)	TMC (%)	CDM (%)
Wireless camera	100	100	100
Mobile camera	80	52	72
Simulated sequences	90	65	75

sets as well. Table 8.1 shows the results, which are best for TMG. For the larger video set of 65 sequences made with the mobile phone 80 % of the slopes is negative, while for the 310 simulated sequences 90 % of the slopes is negative. *The predominantly negative slopes imply that the appearance variation as measured by TMG typically decreases for both the filmed and the simulated approach sequences.* Please note that the entropy of TMG does not only decrease if the obstacle it approaches has no texture, it also decreases if it has one dominant texture or little variation in its texture. Table 8.1 shows that the results are less pronounced for CDM and worse for TMC. This may be due to the color variation not decreasing as much as the texture variation, but it may also be due to difficulties in measuring the color variation. In the remainder of the chapter the focus is solely on TMG, since it provides the best results in terms of the portion of decreasing slopes.

In some sequences the entropy of TMG increases over time. Investigation of these sequences showed that they resemble the one in the bottom row of Fig. 8.2. The simulated camera approaches a book shelf that has objects in it with different, highly detailed textures. These details lead to an increasing entropy as they come into sight. In addition, transparent surfaces such as windows are a problem. The appearance variation depends entirely on the objects behind the transparent surface and not on the surface itself.

The existence of sequences in which the entropy increases and the different offsets of the entropy (see Fig. 8.3) suggest that *the sole use of the appearance variation cue will not lead to perfect obstacle avoidance.*

8.3   Sub-sampling

The idea of sub-sampling is applicable to many computer vision algorithms that
process local image samples. The central, and straightforward, idea is that the algorithms can be made computationally much more efficient by extracting only a subset
of the local image samples, which can for example be extracted at random locations
[8]. In this section, an analysis is presented on the effects that sub-sampling has
on the computational effort (Sect. 8.3.1) and on the performance (Sect. 8.3.2) of the
texton method.

8.3.1   Effect on Computational Effort

The computational effort of the texton method depends on the number of textons n
and the number of extracted samples s. The computational effort c is approximately:

$$c \approx snW + nC, \tag{8.2}$$

where W is the cost of comparing an extracted sample to a texton, and nC is the cost
of calculating the entropy. During execution, n is fixed, but s can be varied freely.
Since the bulk of the computation of the texton method is in the term snW in (8.2),
sub-sampling can lead to speed-ups in the order of hundreds. To see this, think of a
(relatively small) image of 160×120 pixels. Such an image already contains 17,825
possible 5×5 pixel samples. Extracting only 100 local samples leads to a speed-up
of a factor $\sim$178.

Figure 8.4 shows the mean processing times and corresponding standard deviations of a MATLAB implementation of TMG on a data set of 94 images. The
MATLAB-implementation used to obtain these results is available online.[2] The number of samples s is varied from 50 to 2500 with steps of 50. For the experiments,
$n = 30$ and $w = h = 5$. The processing times are measured on a 2.26 GHz laptop.

Figure 8.4 shows that, as expected, the processing times increase linearly with
the number of samples. It gives an idea of the numbers of samples that should be
selected for the task at hand. For example, it is reasonable to state that for obstacle
avoidance with indoor MAVs the execution frequency should at least be 10 Hz, which
corresponds to a processing time of 0.1 s (red dashed line). The figure shows that this
frequency is reached by extracting 250 or fewer samples. In addition, one can see
that one can achieve execution frequencies higher than the frame rate of 30 Hz: using
50 samples results in an execution frequency of 43.7 Hz. Execution of algorithms
at a frequency higher than the frame rate is desirable if one wants multiple vision
processes to run in parallel.

---

[2]http://www.bene-guido.eu/guido/.

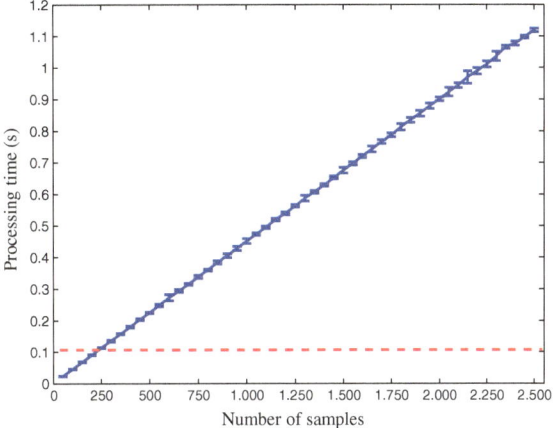

Fig. 8.4 Mean processing times of a MATLAB-implementation of TMG for a number of samples s from 50 to 2500. The error bars indicate the standard deviation

8.3.2 Effect on Performance

The main question that arises when using sub-sampling is what effect it has on a method's performance. In the context of this article, the final performance is how well an MAV can use the estimates from the texton method to achieve successful obstacle avoidance. The number of samples influences the performance successively through:

a. the accuracy of the distribution estimate $\hat{p}$
b. the accuracy of the entropy estimate $H(\hat{p})$
c. the resulting performance of a classifier that uses $H(\hat{p})$ to detect the vicinity of an obstacle
d. the effect this classification has on its use in a controller that brings about obstacle avoidance behaviors.

In the next section, the influence of the number of samples on (a) and (b) are analyzed. The effects on (c) and (d) will be studied in Sects. 8.4 and 8.6, respectively.

8.3.3 Effect on Accuracy

In this section, the relation is studied between the number of samples and the accuracy of the estimates $\hat{p}$ and $H(\hat{p})$. Although the discussion is placed in the context of the appearance variation cue, it is also valid for other applications in which sub-sampling is used for estimating a categorical distribution.

8.3.4 Accuracy of the Distribution Estimate $\hat{p}$

The texton method TMG determines the maximum likelihood estimate of the texton distribution in the image. This distribution is a categorical distribution, and can be fully determined by extracting all samples from the image. Our analysis consists of determining the relation of the number of samples and the expected L1-distance between the maximum likelihood estimate and the actual distribution in the image.

We determine the probabilities for distances between the estimated and actual distributions for the case with replacement. Extracting a fixed number of s samples from random image locations results in texton occurrences $g = \langle g_1, g_2, \ldots, g_n \rangle$. The vector g follows a multinomial distribution and it has the following well-kown probability formula:

$$P(g) = \frac{s!}{g_1! g_2! \ldots g_n!} p_1^{g_1} p_2^{g_2} \cdots p_n^{g_n}, \tag{8.3}$$

with $\sum g_1 + g_2 + \cdots + g_n = s$. For a given number of samples s, this formula allows one to iterate over all possible vectors g while determining the distance d between the estimated distribution $\hat{p} = g/s$ and the actual distribution p. The probability for distance $d = d(\hat{p}, p)$ can then be incremented by $P(g)$. Iterating over all g permits to calculate $P(d)$, the probability distribution of the distances between the estimated and actual distribution.

Although the above method is not elegant, it is tractable for a limited number of samples, since there are also a limited number of possible distances. To illustrate the effects of sub-sampling, we apply the method to the categorical distribution with $n = 6$: $p = \langle 0.5, 0.1, 0.1, 0.05, 0.05, 0.2 \rangle$. Figure 8.5 shows the distribution $P(d)$ (y-axis) for $s = \{50, 100, 150, 200, 250, 300\}$ (x-axis), where illuminance represents high (white) to low (black) probabilities. The black solid line indicates the mean distance to the actual distribution and the white dashed line the 95th percentile.

Two main observations can be made from Fig. 8.5. First, as to be expected, the mean of the distribution gets closer to the actual distribution as the number of samples increases. Second, this effect obeys the law of diminishing returns, so that *a modest number of samples already reduces the probability for 'large' distances considerably.*

When the distribution is known, $P(d)$ can be determined to provide certainty bounds to the distance between the estimated and actual distribution. For example, for the categorical distribution with replacement given above, $P(d < 0.20 \wedge s = 150) = 0.95$. The problem is of course that the actual distribution is not available. Still, we can limit the 95 % certainty bound from above by assuming the worst case scenario, which occurs when the entropy of the actual distribution is highest. Figure 8.6 shows the mean distances and the 95th percentiles for different distributions. The distributions have $n = 6$ bins and different entropies mentioned in the figure (the colors range from green to red, indicating low to high entropy). The distances are highest for the distribution with maximal entropy, $H = 2.585$.

The relation between entropy and the mean L1-distance could be particularly relevant to the appearance variation cue, since it uses the entropy $H(\hat{p})$ as a measure

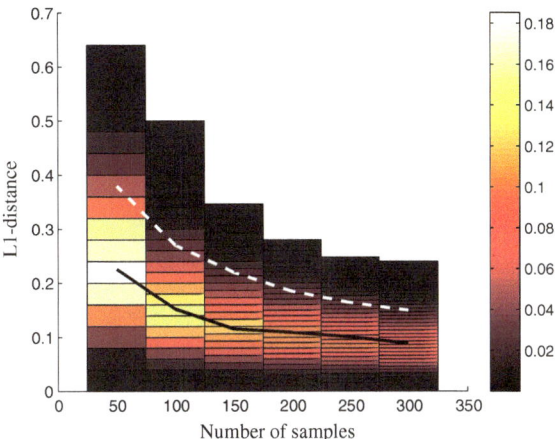

Fig. 8.5 Relation between the number of samples used in sub-sampling and the accuracy of the estimate. The y-axis shows the L1-distance between the estimated and the true distribution, the x-axis shows the number of samples, $s = \{50, 100, 150, 200, 250, 300\}$. The color map represents the probability, where bright yellow indicates high probability and dark red a low probability. The black solid line indicates the mean distance from the actual distribution, the white dashed line the 95th percentile of the distribution

Fig. 8.6 Relation between the entropy H and the mean L1-distances (*solid lines*) and the 95th percentile (*dashed lines*) for different numbers of samples. The color of the lines ranges from green (low entropy) to red (high entropy). The highest entropy leads to the least accurate estimates

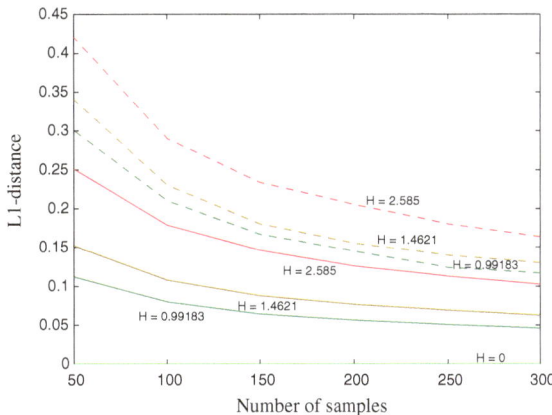

of obstacle proximity. Near-collisions are associated with low entropy distributions, which suffer the least from sub-sampling in terms of L1-distance.

8.3.5 Accuracy of the Entropy Estimate $H(\hat{p})$

Another aspect of the accuracy of TMG with sub-sampling is the effect of the number of samples on the error $H(p) - H(\hat{p})$. Figure 8.7 shows the true entropy (black line) and the average entropy estimates for different numbers of samples for a single

Fig. 8.7 Estimated entropy
over time for a single
approach sequence. The
black line is the entropy of
the true distribution. The
differently colored lines
show the mean estimated
entropy for $s =$
$\{5, 10, 50, 100, 200, 400\}$,
colored from *green* to *red*.
The standard deviations of
the estimates are shown by
the error bars

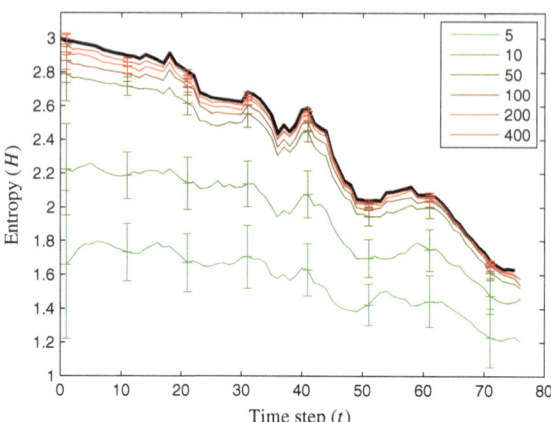

approach sequence of an obstacle ($n = 30$). The studied numbers of samples are
$s = \{5, 10, 50, 100, 200, 400\}$, indicated with colors going from green to red. The
standard deviations of the different numbers of samples are shown by the error bars.

There are three observations to be made from this figure. First, the figure clearly
shows the well-known fact that the maximum likelihood estimate $\hat{p}$ on average leads
to underestimation of the entropy: $E[H(\hat{p})] \leq H(p)$, c.f. [13,18].[3] The lower the
number of samples, the larger this estimation bias is. Second, as to be expected, the
standard deviation of the estimates is larger for lower numbers of samples. With
respect to the appearance cue, it can be remarked that already for as few a number of
samples as 50, the standard deviations are much smaller than the change in entropy
over time. This implies that the entropy measurements will be reliably lower when
closer to an obstacle, which allows the use of a threshold for determining obstacle
proximity. Please note that in the context of the appearance variation cue the magnitude of the estimation bias $E[H(p) - H(\hat{p})]$ is less relevant. However, the bias
does imply that the threshold value should be changed if the number of samples s
is changed. Third, the estimation bias seems to be higher at the start of the run than
at the end of the run. The reason for this is that the estimation bias has a similar
relation to $H(p)$ as the L1-distance: a higher $H(p)$ leads to higher error. Figure 8.8
shows the relation between the estimation bias $E[H(p) - H(\hat{p})]$ and the number of
samples s for distributions with a different (true) entropy $H(p)$, with $n = 30$ bins.
The colors of the lines go from green (low entropy) to red (high entropy).

In summary, for $n = 30$ the negative effect of sub-sampling on the accuracy of
$\hat{p}$ and $H(\hat{p})$ seems limited already for a number of samples in the range of 50, 100.
Moreover, with respect to the appearance variation cue, near-collisions (associated
with low entropy distributions) suffer the least from sub-sampling.

---

[3]In fact, using the Delta method, in [1] the following formula was obtained: $E[H(\hat{p})] \sim H(p) -$
$\frac{n-1}{2s}$.

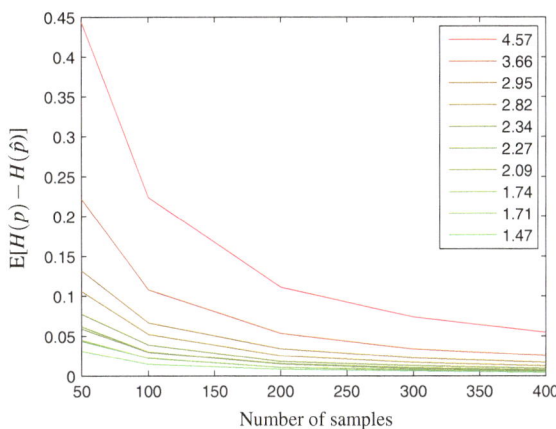

Fig. 8.8 The estimator bias $E[H(p) - H(\hat{p})]$ for different numbers of samples for distributions p with various entropies. The entropies range from 1.47 to 4.57, indicated with the colors from *green* to *red*

8.4   Classification Experiments

In Sect. 8.2, we indicated three reasons why the appearance variation cue could be complementary to the use of optic flow for obstacle avoidance. In this section optic flow is applied to video sequences in order to calculate the time-to-impact τ. We investigate a task in which the appearance variation and/or the optic flow are used to classify whether the time-to-impact[4] is larger or smaller than a desired detection time τ^*.

Two types of video sequences are used: (1) the 65 mobile phone videos discussed in Sect. 8.2, and (2) 19 in-flight videos made with the camera on board the DelFly II (shown in Fig. 8.9). For the in-flight videos, a human pilot steered the DelFly II from approximately 3 m towards walls and other large obstacles inside different office rooms. The difference between the DelFly sequences and the video sequences made by hand include the more complex dynamics of the camera's movements and noise introduced by the DelFly's motor, which is powered by the same battery as the camera.

The method TMG is applied to the images with parameter settings $s = 100$ and $n = 30$. Image samples are extracted at uniformly distributed locations (with replacement).

The remainder of this section is structured as follows. In Sect. 8.4.1 the optic flow method is explained. In Sect. 8.4.2 the classification performances of the methods are evaluated.

---

[4]Please note that the appearance variation cue does not directly depend on the time-to-impact, but on the distances to obstacles in view. However, by assuming a constant velocity the time-to-impact and distance to the imminent obstacle are linearly related. For this reason, both methods can be compared on the time-to-impact classification task.

Fig. 8.9 The flapping wing
MAV DelFly II, with a wing
span of 28 cm and a weight
of 16 grams. It carries two
cameras onboard that
transmit images via an
analog channel. In the
current experiments, only the
forward-looking camera is
used

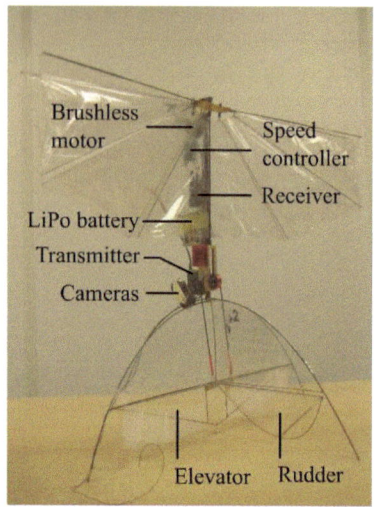

8.4.1 Optic Flow for Calculating Time-to-impact

The context of obstacle avoidance with light-weight MAVs makes computational efficiency and robustness to noisy images the main priorities for the optic flow algorithm. For this reason, simplicity of the optic flow algorithm is preferred over detailed information on the environment. The implementation of the algorithm has two parts: (1) finding and tracking feature points to determine several optic flow vectors between two images, and (2) determining τ on the basis of the vectors.

For the first part, the openCV library is used to find corner-like feature points in the image and track them over time [2, 17, 20]. Specifically, the algorithm calculates the following matrix G at every image location (x, y):

$$G(x, y) = \begin{bmatrix} I_x^2(x, y) & I_x(x, y)I_y(x, y) \\ I_x(x, y)I_y(x, y) & I_y^2(x, y) \end{bmatrix}, \tag{8.4}$$

where $I_x(x, y)$ and $I_y(x, y)$ are the horizontal and vertical image gradient at image position (x, y), respectively. The larger the smallest eigenvalue $\lambda_s(x, y)$ of this matrix, the larger the probability that a corner is present at location (x, y). The algorithm finds the maximal such eigenvalue in the image and then retains the locations at which $\lambda_s(x, y) > c \max_{x,y}(\lambda_s(x, y))$, where $c \in [0, 1]$. Although there are many other approaches of finding suitable features and extracting optic flow (c.f., [4, 10–12, 25]), the openCV implementation is computationally efficient and works well for both large and small flows between subsequent images. Hence, it lends itself well for applications with MAVs, e.g., [5].

The second part is performed as follows. It is assumed that the camera is moving straight towards a flat surface, while rotation is assumed to be limited. The Point of Expansion (PoE) is estimated with the least-squares method described in [22]. Since considerable amounts of noise are expected, the method is first applied to all optic flow vectors and then repeated on the inliers. Subsequently, the optic flow vectors are used to determine the distance from the old location (x_t, y_t) and the new location (x_{t+1}, y_{t+1}) to the PoE (x_e, y_e); $d_{e,t}$ and $d_{e,t+1}$ respectively. The difference in distance to the PoE is $\Delta d_{e,t}$. Each optic flow vector leads to one estimate $\hat{\tau}$:

$$\hat{\tau} = d_{e,t}/\Delta d_{e,t} \qquad (8.5)$$

Since it is assumed that there is one flat surface and the optic flow vectors are noisy, the final estimate τ_{OF} is taken to be the median of the resulting $\hat{\tau}$-distribution. The uncertainty of τ_{OF} can be captured with the standard deviation $\sigma(\tau_{OF})$. Despite the strong assumptions, this straightforward method works rather well in practice.

8.4.2   Classification Performance

After determining TMG's entropy, τ_{OF}, and $\sigma(\tau_{OF})$ on the set of videos, we investigate the classification performances of different combinations of methods. The task is to classify a time step t as positive when $\tau \leq \tau^*$. The following logical expressions are used for classification: '$\tau_{OF} < \vartheta(\tau_{OF})$', 'TMG $< \vartheta$(TMG)', '$\tau_{OF} < \vartheta(\tau_{OF}) \wedge \sigma(\tau_{OF}) < \vartheta(\sigma(\tau_{OF}))$', '$\tau_{OF} < \vartheta(\tau_{OF}) \vee$ TMG $< \vartheta$(TMG)', and '$(\tau_{OF} < \vartheta(\tau_{OF}) \wedge \sigma(\tau_{OF}) < \vartheta(\sigma(\tau_{OF}))) \vee$ TMG $< \vartheta$(TMG)'. The rationale behind this last expression is that the optic flow estimate should only be trusted when it is accurate enough, with the entropy of TMG to ensure detection if this is not the case. For brevity, the thresholds ϑ will be omitted from here on.

By varying the thresholds in the above expressions, a Receiver Operator Characteristic (ROC) curve [9] can be made that represents the trade-off between the true positive ratio (proportion of time steps with $\tau < \tau^*$ that are classified as positive) and the false positive ratio (proportion of time steps with $\tau > \tau^*$ that are classified as positive). Remark that by varying the threshold of more than one variable, one obtains many points in the two-dimensional space of the true positive and false positive ratios. These points will not necessarily lie on a curve. In the experiments, the best threshold combinations are retained leading to one ROC-curve. Since this procedure essentially involves the selection of threshold settings to achieve a high true positive ratio and low false positive ratio, it introduces a risk of overfitting. Therefore, the best threshold combinations are determined on the first half of the available data and then applied to the second half.

Starting with the results on the mobile phone videos, Fig. 8.10 shows the ROC-curves for the methods τ_{OF} (dotted), TMG (thin solid), $\tau_{OF} \wedge \sigma(\tau_{OF})$ (dash-dotted), $\tau_{OF} \vee$ TMG (dashed), and $(\tau_{OF} \wedge \sigma(\tau_{OF})) \vee$ TMG (thick solid) for $\tau^* = 40$ (expressed in frames to impact). The x-axis represents the ratio of false positives, while the y-axis represents the ratio of true positives. The higher the curve, the better. The combination of all three variables obtains the best results; it has the highest ratio of true positives at almost all rates of false positives.

The quality of the classifier can be expressed by the Area Under the Curve (AUC). Table 8.2 shows the AUC-values for the different methods for different τ^*. A bold setting indicates the best method. In four out of five cases, a combination containing

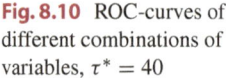

Fig. 8.10 ROC-curves of different combinations of variables, $\tau^* = 40$

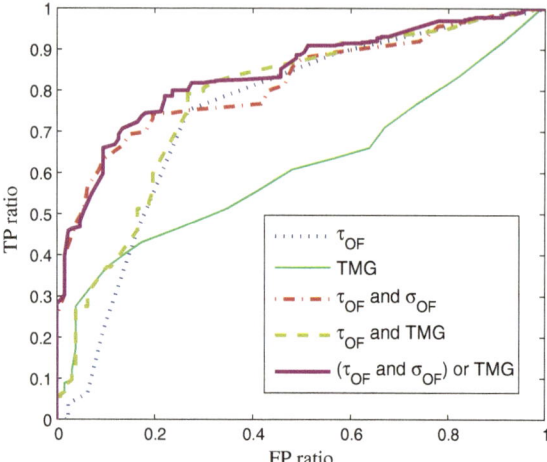

Table 8.2 AUC for combinations of τ_{OF}, $\sigma(\tau_{OF})$, and TMG. Bold indicates the highest AUC. Video sequences made by hand

τ^*	τ_{OF}	TMG	$\tau_{OF} \wedge \sigma(\tau_{OF})$	$\tau_{OF} \vee$ TMG	$(\tau_{OF} \wedge \sigma(\tau_{OF})) \vee$ TMG
10	0.836	0.881	0.769	0.876	**0.897**
20	0.893	**0.976**	0.848	0.963	0.877
30	0.926	0.927	0.899	**0.942**	0.936
40	0.908	0.819	0.896	0.867	**0.944**
50	0.887	0.733	0.880	0.769	**0.974**

both TMG and τ_{OF} works better than either measure alone. Importantly, these results have been obtained while TMG only extracted 100 samples from each image.

The same types of results are shown for the video sequences made by flying the DelFly II. Table 8.3 contains the AUC values for all different τ^*. The best performances are obtained by combinations of τ_{OF} and TMG. Figure 8.11 shows the ROC curves for $\tau^* = 40$. For this threshold, the lowest FP ratios for a TP ratio smaller than 0.91 are obtained either by TMG alone or by TMG together with τ_{OF}. For higher TP ratios, first the combination of TMG with $\sigma(\tau_{OF})$ and τ_{OF} gives the best results, and then the combination of τ_{OF} and $\sigma(\tau_{OF})$.

The most remarkable difference between the results on the DelFly sequences and the manually made sequences is that the optic flow method τ_{OF} is constantly outperformed by TMG. The degradation in performance of the optic flow method may be caused by increased image noise or by the more complex camera motion. Although the pilot attempts to steer the DelFly straight towards an obstacle, slight deviations of this path occur because of external factors such as drafts. These different movements and the three-dimensional structure of the world violate the assumptions underlying the optic flow implementation.

Table 8.3 AUC for combinations of τ_{OF}, $\sigma(\tau_{OF})$, and TMG. Bold indicates the highest AUC. DelFly II sequences

τ^*	τ_{OF}	TMG	$\tau_{OF} \wedge \sigma(\tau_{OF})$	$\tau_{OF} \vee$ TMG	$(\tau_{OF} \wedge \sigma(\tau_{OF})) \vee$ TMG
10	0.817	0.844	0.801	0.850	**0.860**
20	0.752	0.848	0.739	**0.849**	0.805
30	0.745	0.934	0.753	**0.935**	0.914
40	0.718	0.916	0.768	**0.918**	0.901
50	0.661	0.853	0.730	0.854	**0.869**

Fig. 8.11 ROC-curves of different combinations of variables, $\tau^* = 40$, for the video sequences made with the flying DelFly II

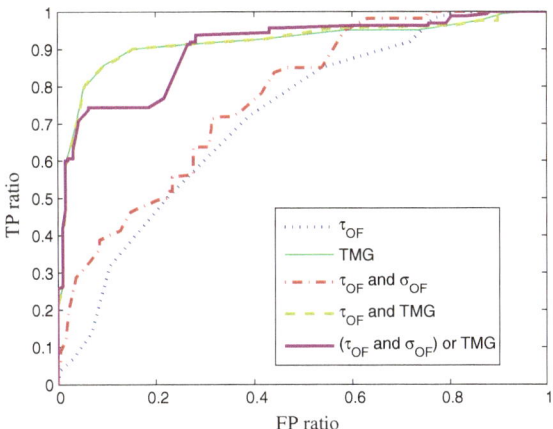

A more evident cause for TMG outperforming τ_{OF} is that *the high-frequent flapping movements of the DelFly in combination with the line-by-line recording of the video images leads to image deformations.* The deformations can be more or less severe, depending on the flapping frequency and other factors such as the physical connection between the camera and the robot body. Figure 8.12 shows two subsequent images made with the camera onboard the DelFly II. These images are particularly affected by the flapping movements: the legs of the person are thinner and curved in the left image. As a consequence of these unpredictable deformations, images such as those in Fig. 8.12 do not comply anymore with the linear camera model assumed by optic flow algorithms. For this reason, we do not expect other optic flow algorithms to perform much better on the DelFly images. Please note that currently there are no global shutter cameras of a scale and weight suitable for the DelFly II or DelFly Micro. Such cameras may become available in the future though, leading to an improvement of the performance of optic flow algorithms on these flapping wing MAVs.

Fig. 8.12 Two subsequent images in a video made onboard the DelFly II. In addition to the well-known image deformations by the lens, there are deformations caused by the combination of the flapping frequency of the DelFly and the line-by-line recording of the camera. As a consequence, the images do not conform to a linear camera model. This degrades the performance of optic flow methods for this specific application

8.5   Simulated Avoidance Experiments

The classification results in the last section show that the measurements of optic flow and the texton method can be complementary. The results are based on observations of all sequences and time steps mixed together. In addition, in the classification experiments, there is no relation between the behavior of the camera and the measurements. For these reasons, the question remains what the classification results mean in terms of obstacle avoidance behavior. In this section, the behavior of a simple controller is investigated. Since testing of many different parameter settings is infeasible on the real robot, these experiments are performed in simulation.

8.5.1   Experimental Setup

The goal of the simulation experiments is to determine whether τ_{OF} and TMG are also complementary for achieving successful avoidance behavior. The experiments focus on a straightforward controller that flies straight until it detects an obstacle. In that case, it turns to the right for a fixed number of time steps, during which it does not evaluate its inputs. The dynamics employed in these experiments are very limited: the simulated DelFly does not change in height and its trajectory is not influenced by external factors.

The expression used for evaluating the vicinity of an obstacle is '$(\tau_{OF} < \vartheta(\tau_{OF}) \wedge \sigma(\tau_{OF}) < \vartheta(\sigma(\tau_{OF}))) \vee TMG < \vartheta(TMG)$'. In the experiments a grid search is performed on the three thresholds. As a result of the grid search, all other expressions studied in the previous section will arise as special cases. For example, the expression $\tau_{OF} < \vartheta(\tau_{OF}) \vee TMG < \vartheta(TMG)$ arises when $\vartheta(\sigma(\tau_{OF})) > \max(\sigma(\tau_{OF}))$.

For each threshold setting, the simulated DelFly II is evaluated on an obstacle avoidance task in the simulator. Figure 8.13 shows an image of the simulated DelFly in its simple environment: a room with four walls, a floor and a ceiling. While the floor

Fig. 8.13 The simulation environment with the simulated DelFly II. The floor and ceiling are both white, so that they do not contain any texture. The texture on the walls is varied by filling their surface to a certain proportion with photographs and paintings

and ceiling are white, the texture on the walls can vary. By automatically changing the proportion of the walls that is covered with textured objects, the relation between the amount of texture and the performance of the different threshold settings can be evaluated. Figure 8.14 shows three different walls, with 5, 15, and 25 % of their surface covered by textured objects.

For comparison of the simulated camera images with real camera images, we measure the texture in the image with the help of the corner-measure used by the optic flow algorithm. The texture measure is the average of the minimal eigenvalues of the matrix mentioned in Sect. 8.4.1:

$$\overline{\lambda}_s = \frac{\sum_{x=1}^{width} \sum_{y=1}^{height} \lambda_s(x, y)}{width \times height}, \tag{8.6}$$

where $\lambda_s(x, y)$ is the smallest eigenvalue of the matrix $G(x, y)$ defined in Eq. 8.4. The larger $\overline{\lambda}_s$, the more texture is present in the image. The simulated camera images in environments with the texture proportions mentioned in Fig. 8.14 correspond on average to $\overline{\lambda}_s = 0.00035, 0.00048$, and 0.00071. For comparison, in the video sequences made by hand (see Sect. 8.2.1) and with the DelFly (Sect. 8.4) $\overline{\lambda}_s = 0.00023$ and 0.00052, respectively.

In the experiments, the thresholds are varied with five steps in the intervals: $\vartheta(\tau_{OF}) \in [0, 80]$, $\vartheta(\sigma(\tau_{OF})) \in [20, 100]$, and $\vartheta(TMG) \in [0.0, 2.4]$. For each setting, the DelFly has to fly in six different environments with textures ranging from 0 to 25 % (with steps of 5 %). In each environment the DelFly flies around until either (i) two minutes have passed, or (ii) the DelFly crashes into a wall.

After testing one threshold setting, three statistics are calculated to evaluate the performance: the Avoidance (A), the Rudder deflection (R), and the Exploration (E).

Fig. 8.14 Three of the walls used in the simulator, generated to achieve a texture surface proportion of 0.05, 0.15, and 0.25

Performance measure A is measured as:

$$A = \frac{1}{6} \sum_{i=1}^{6} \frac{f_i}{F}, \tag{8.7}$$

where i is the iterator over the different textures of the environment, f_i is the number of time steps flown, and F the maximal number of time steps representing two minutes. The measure R captures how much the simulated DelFly is turning. A controller that turns continuously is not suitable for autonomous flight, since it does not lead to any exploration of the environment and may in the real world still lead to crashes. R is measured as:

$$R = \frac{1}{6} \sum_{i=1}^{6} \left(\frac{1}{f_i} \sum_{t=1}^{f_i} (1 - r) \right), \tag{8.8}$$

where $r \in [0, 1]$ is the rudder deflection, which is 0.8 when turning. The $(1 - r)$ term has as a consequence that a high R corresponds to a good performance (few turns), while a value of 0.2 means that the DelFly is continuously turning.

The goal of the performance measure E is to evaluate how well the DelFly has explored its environment. To determine the exploration, the environment is divided in $100 \times 100 = 10{,}000$ grid cells. A cell is explored in run i ($e_i(x, y) = 1$), if there is at least one position of the DelFly that falls within the cell. Else $e_i(x, y) = 0$. The equation for E then is:

$$E = \frac{1}{6} \sum_{i=1}^{6} \left(\frac{1}{10000} \sum_{x=1}^{100} \sum_{y=1}^{100} e_i(x, y) \right). \tag{8.9}$$

We would like to remark that even in a perfect run with only necessary turns, the value of E has a low value: it will not reach 1 in two minutes simulation time.

8.5.2 Results

The grid search over three parameters leads to a 4-dimensional matrix per performance measure. In Fig. 8.15 we show slices through these matrices where

$\vartheta(\sigma(\tau_{OF})) = 80$. This value is selected, because it contains the maximum for the exploration measure. The three plots represent the avoidance, rudder, and exploration performance values.

The maximum exploration performance is reached with the parameter setting $\vartheta(\tau_{OF}) = 80$, $\vartheta(\sigma(\tau_{OF})) = 80$, $\vartheta(TMG) = 2.4$. This best performance is achieved by using a combination of both TMG and the optic flow method. The figure also shows that parameter settings ignoring either method ($\vartheta(TMG) = 0$ or $\vartheta(\tau_{OF}) = 0$) obtain lower performances.

The reason for the better performance of the combined methods can be seen by investigating the values of TMG and τ_{OF} when flying around in differently textured environments. Figure 8.16 shows for different amounts of texture $\overline{\lambda_s}$ the proportion of time steps during which the values of $\sigma(\tau_{OF})$, τ_{OF}, and TMG are below their respective thresholds.

Two main observations can be made from this figure that confirm the complementarity between TMG and τ_{OF}. The first and most remarkable observation to be made from Fig. 8.16 is that τ_{OF} is always under its threshold $\vartheta(\tau_{OF})$ at the lowest amount of texture. The explanation for this is that in our implementation the absence of texture leads to random large optic flow vectors. This leads to small (and sometimes even negative) values of τ_{OF}. Hence, for the lowest texture the comparison of

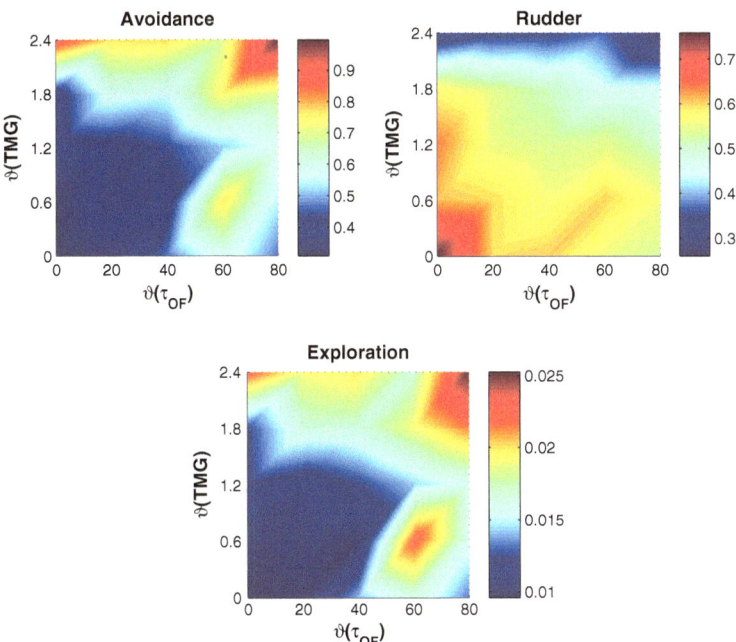

Fig. 8.15 Avoidance, rudder, and exploration performance measures for $\vartheta(\sigma(\tau_{OF})) = \vartheta(\sigma(\tau_{OF})) = 80$. The behaviors with highest performance are obtained with a combination of the appearance variation cue (TMG) and the optic flow method

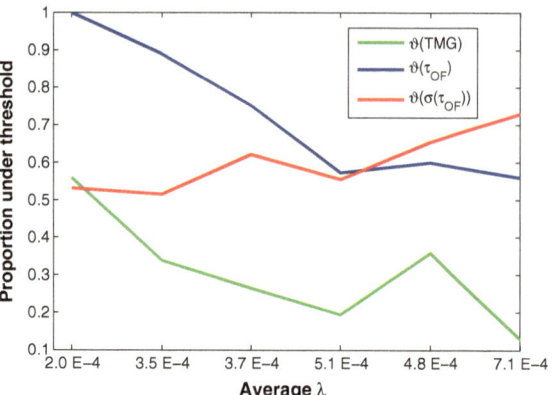

Fig. 8.16 Proportion of time steps during which the values of $\sigma(\tau_{OF})$, τ_{OF}, and TMG are below their respective thresholds versus the texture measure $\overline{\lambda_s}$

the estimated optic flow to its threshold is not informative. As the texture increases, τ_{OF} becomes more informative: the uncertainty of τ_{OF} as measured by the standard deviation decreases on average. This can be seen in the figure, since $\sigma(\tau_{OF})$ is more often under its threshold $\vartheta(\sigma(\tau_{OF}))$. The second observation is that TMG is less often under its threshold $\vartheta(TMG)$ as the texture increases. For higher textures, the controller relies more and more on optic flow.

8.6 Real-World Avoidance Experiments

The final experiment involves the use of both the appearance variation cue and optic flow to control the DelFly II in order to verify whether the detection performance is good enough for successful obstacle avoidance.

During the experiment, the onboard video images are sent to a ground station. While the vision algorithms are currently running offboard, their efficiency remains of uttermost importance in order to control small MAVs with fast dynamics and to have a chance of being implemented onboard in the near future. The video frame rate is 30 Hz. The frames are down-sized to 160×120 images and processed by both the optic flow algorithm running at 30 Hz and the texton method running at 13 Hz ($n = 30$ and $s = 100$). The execution frequencies have been measured on a dual-core 2.26 GHz laptop with all ground station software running at the same time.

While the height of the DelFly is controlled by a human pilot, the rudder is entirely controlled by the ground station. The algorithm for the rudder control is illustrated with a flowchart in Fig. 8.17. In case of a collision detection, the control of the rudder is straightforward: it performs a sequence of open loop commands for $\sim$1.5 seconds. The sequence is always executed fully before new commands are allowed. When not executing a turn, a collision is detected if $(\tau_{OF} >= 0 \wedge \tau_{OF} < 2 \wedge \sigma_{OF} < 2) \vee TMG <$ 2.2. In order to prevent the DelFly from getting into a downward spiral, collision

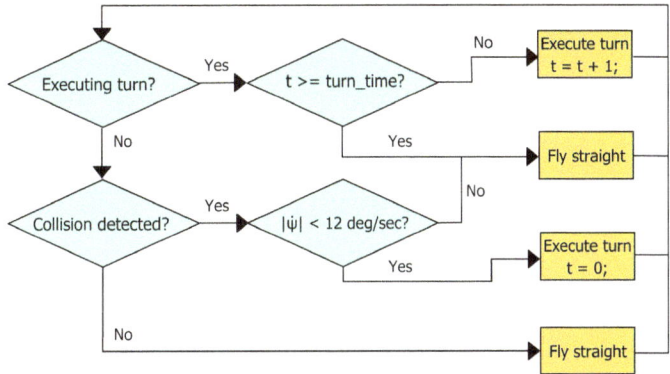

Fig. 8.17 Flowchart of the obstacle avoidance algorithm controlling the DelFly II during the avoidance experiments

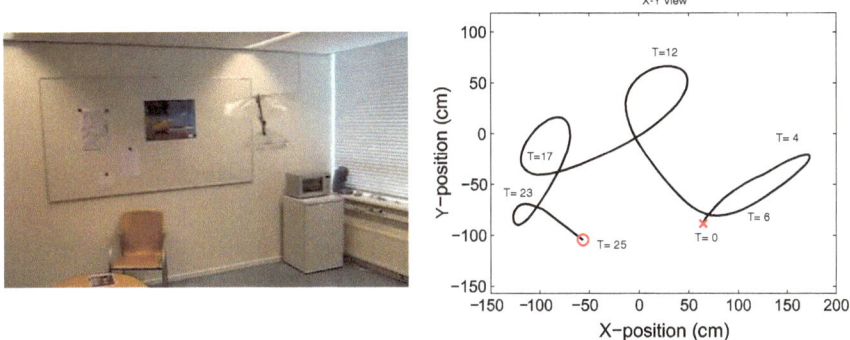

Fig. 8.18 *Left* picture of the office environment in which the experiment took place. *Right* trajectory of the DelFly during one of the experiments

detections only result in turning behavior if the absolute yaw rate ($|\dot{\psi}|$) is lower than 12°/sec. The yaw rate is estimated by means of the median horizontal flow f_x as follows: $\dot{\psi} = 2f_x/\text{FOV}$, where FOV is the horizontal field of view.

The DelFly successfully avoided obstacles in two different office spaces. Figure 8.18 shows one of these spaces on the left. The right part of the figure shows the trajectory of one of the experiments, from the launch of the DelFly (X) to when it is caught (O). The DelFly's x, y-position is plotted over time. The borders indicate the large obstacles such as walls in the room, and hence delimit the flyable area. The trajectory shows that the DelFly detects all obstacles on time, while not performing an excessive number of avoidance maneuvers. As to be expected, not every open loop turn has an equal effect. One factor playing a role is that subsequent open loop turns tend to result in sharper turns, but external factors such as drafts also have a big influence on the dynamics of the 16-gram DelFly.

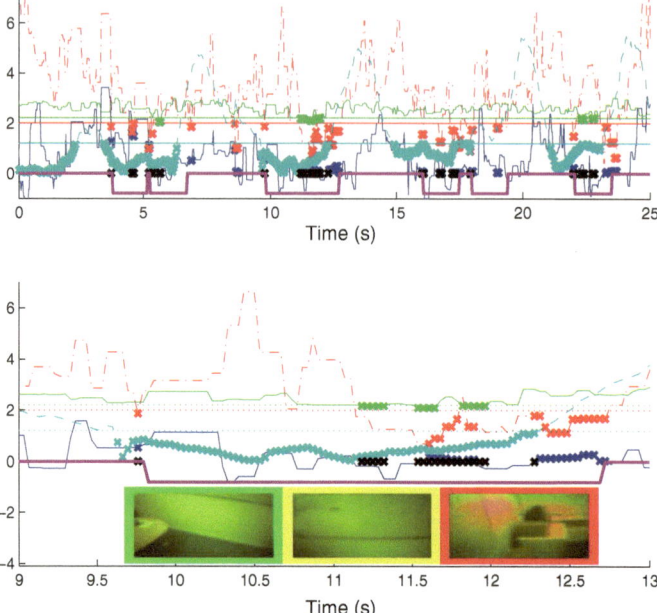

Fig. 8.19 *Top* all relevant variables during the entire experiment. *Bottom* two obstacle avoidance maneuvers, the first triggered by optic flow and the second by the appearance variation. See the text for further details

The top plot of Fig. 8.19 shows the variables computed on the basis of the onboard images during the entire experiment. The x-axis shows the time in seconds. The y-axis shows the values of the variables relevant to obstacle avoidance: τ_{OF} (dark blue solid), σ_{OF} (red dash dotted), TMG (green solid), and the absolute yaw rate $|\dot{\psi}|$ scaled by a factor 0.1 for visualization purposes (light blue dashed). The corresponding thresholds are shown with dotted lines in the same colors. Finally, the plot also shows the command c sent to the rudder (purple solid), for flying straight ($c = 0$) or making a turn ($c = -0.8$). Colored crosses in the figure show when the variables satisfy their conditions for an obstacle detection. The black crosses at $y = 0$ show when a collision is detected and $|\dot{\psi}|$ allows for a new open loop turn.

In the bottom plot of Fig. 8.19 the focus is on a time interval in which the DelFly performs two subsequent open loop turns. The first turn is the consequence of a false positive detection by the optic flow measurements. The turn directs the DelFly towards a wall, necessitating collision avoidance. The wall has little texture and is first detected on the basis of the appearance variation cue, leading to a timely turn away from the wall. The onboard image at the moment of detection is shown in the figure with a yellow border. Videos and additional material concerning the experiments are available at http://www.bene-guido.eu/guido/.

8.7 Discussion

The empirical results in the previous sections have shown that the appearance variation cue can be a useful complement to the time-to-impact as determined by optic flow algorithms. Still, many aspects of the novel visual cue require further study. One aspect is the way of measuring the appearance variation. For example, the current study has not focused on thoroughly optimizing the number of textons and the dictionaries of TMG and TMC, which may well be the reason for TMC's low performance in Sect. 8.2.

In addition, it is necessary to identify the situations for which the algorithm is most suitable. This can be related to both the robotic platform and the environment. For example, the classification experiments show that the appearance variation cue has a strong added value for the images made with the DelFly II. The reason for this seems to be the deformation of the images by the flapping movements. Other aspects can also influence the usefulness of the appearance variation cue. A small camera with infinity focus will result in increased blur when close to an obstacle, smearing out any detailed texture. A camera platform that vibrates continuously will also result in larger (motion) blur when closer to an obstacle. Both these effects are advantageous for the appearance variation cue.

We have focused on indoor, man-made environments. These are typically less textured than outdoor, natural environments. Indeed, optic flow is often more successful outdoors than indoors. As a preliminary test on outdoor environments, we measure the appearance variation of 65 videos made with a mobile phone. In the videos various obstacles are approached such as trees, bushes, and outside walls. The proportions of negative slopes (determined as in Sect. 8.2) are: for TMG 57.0 %, for TMC 46.1 %, and for CDM 78.5 %. A preliminary analysis shows that the texture variation as measured by TMG does not decrease reliably, since obstacles such as trees have an inhomogeneous detailed texture. On the other hand, the color variation as measured by CDM does decrease, since there is less variation in a tree bark's color alone than in the environment surrounding it. Because the percentage of decreasing entropies for CDM outdoors is lower than the percentage for TMG indoors, it seems that the appearance variation cue may be more suited for indoor environments. Hence, optic flow and the appearance variation cue may be complementary concerning the type of environment in which they function best.

8.8 Conclusions

We conclude that in indoor environments the appearance variation cue is a useful complement to optic flow for obstacle avoidance. The experiments indicate that in ~80–90 % of cases, the appearance variation as measured by the texton method TMG decreases towards impact. The variation can increase in cases where the obstacle has a detailed texture. This finding implies that the appearance cue cannot be used in isolation to detect all obstacles. Nonetheless, it can be combined successfully with optic flow: the AUC of a classifier using both cues is higher than that of either cue alone. This effect seems larger for more degraded images. In the closed-loop experiments the combination of both visual cues allowed the flapping wing MAV DelFly II to successfully detect and avoid both textured and texture-poor obstacles.

With respect to sub-sampling, our main conclusion is that it can make the extraction of the appearance variation cue fast enough for use in indoor flight. Sub-sampling improves the computational efficiency of extracting the cue by a factor ~ 100, while retaining an acceptable accuracy. Analysis showed that (1) increasing the number of samples obeys the law of diminishing returns regarding the accuracy, and (2) given a number of samples the accuracy of the sub-sampling method is best when the entropy of the true distribution is low.

Acknowledgments   This chapter is partly based on [6,7].

References

1. G.P. Basharin, On a statistical estimate for the entropy of a sequence of independent random variables. Theor. Probab. Appl. **4**(3), 333–336 (1959)
2. J-Y. Bouguet, Pyramidal implementation of the Lucas Kanade feature tracker. Description of the algorithm (2000)
3. V. Bruce, P.R. Green, M.A. Georgeson, *Visual Perception: Physiology, Psychology and Ecology*. (Psychology Press, Routledge, 2003)
4. A. Bruhn, J. Weickert, C. Feddern, T. Kohlberger, C. Schnorr, Real-time optic flow computation with variational methods, in *CAIP 2003, LNCS 2756* (2003), pp. 222–229
5. J. Conroy, G. Gremillion, B. Ranganathan, J.S. Humbert, Implementation of wide-field integration of optic flow for autonomous quadrotor navigation. Auton. Robot. **27**(3), 189–198 (2009)
6. G.C.H.E. de Croon, E. de Weerdt, C. de Wagter, B.D.W. Remes, R. Ruijsink, The appearance variation cue for obstacle avoidance, in *ROBIO* (2010)
7. G.C.H.E. de Croon, E. de Weerdt, C. de Wagter, B.D.W. Remes, R. Ruijsink, The appearance variation cue for obstacle avoidance. IEEE Trans. Robot. **28**(2), 529–534 (2012)
8. G.C.H.E. de Croon, C. De Wagter, B.D.W. Remes, R. Ruijsink, Sub-sampling: real-time vision for micro air vehicles. Robot. Auton. Syst. **60**(2), 167–181
9. T. Fawcett, An introduction to roc analysis. Pattern Recogn. Lett. **27**, 861–874 (2006)
10. D.J. Fleet, A.D. Jepson, Stability of phase information. IEEE Trans. Pattern Anal. Mach. Intell. **15**(12), 1253–1268 (1993)
11. T. Gautama, M.M. Van Hulle, A phase-based approach to the estimation of the optical flow field using spatial filtering. IEEE Trans. Neural Networks **13**(5), 1127–1136 (2002)
12. B.K.P. Horn, B.G. Schunck, Determining optical flow. Artif. Intell. **17**, 185–203 (1981)
13. B.M. Jedynak, S.M. Khudanpur, Maximum likelihood set for estimating a probability mass function. Neural Comput. **17**(7), 1508–1530 (2005)
14. L. Kaufman, *Sight and Mind* (Oxford University Press, New York, 1974)
15. T. Kohonen, *Self-Organizing Maps*. (Springer, Berlin, 2001)
16. H.C. Longuet-Higgins, K. Prazdny, The interpretation of a moving retinal image. Proc. R. Soc. Lond. B **208**, 385–397 (1980)

17. B.D. Lucas, T. Kanade, An iterative image registration technique with an application to stereo vision, in *Proceedings of Imaging understanding workshop* (1981), pp. 121–130
18. T. Schuermann, Bias analysis in entropy estimation. J. Phys. A Math. Gen. **37**, 295–301 (2004)
19. C.E. Shannon, A mathematical theory of communication. Bell Syst. Tech. J. **27**(379–423), 623–656 (1948)
20. J. Shi, C. Tomasi, Good features to track, in *CVPR* (1994)
21. R.T. Surdick, T.D. Davis, T. Elizabeth, R.A. King, G.M. Corso, A. Shapiro, L. Hodges, K. Elliot, Relevant cues for the visual perception of depth: Is where you see it where it is? in *Human Factors and Ergonomics Society Annual Meeting Proceedings, Visual Performance* (1994), pp. 1305–1309(5)
22. N. Takeda, M. Watanabe, K. Onoguchi, Moving obstacle detection using residual error of FOE estimation, in *IROS* (1996), pp. 1642–1647
23. L.J.P. van der Maaten, An introduction to dimensionality reduction using matlab, micc 07-07. Technical report, Maastricht University, The Netherlands (2007)
24. M. Varma, A. Zisserman, Texture classification: are filter banks necessary? in *(CVPR 2003)*, vol 2 (2003), pp. 691–698
25. T. Zhang, H. Wu, A. Borst, K. Kuhnlenz, M. Buss, An fpga implementation of insect-inspired motion detector for high-speed vision systems, in *2008 IEEE International Conference on Robotics and Automation Pasadena, CA, USA* (2008), pp. 335–340

Optical Flow Based Turning Logic

9

Abstract

The DelFly does not use a map of its environment to achieve successful obstacle avoidance. Instead, it uses more direct visual cues to detect obstacles and to turn in the right direction. In this chapter, we investigate how successful such a direct avoidance strategy can be. We introduce a turning logic for the DelFly and test it both in simulation and on the real DelFly II platform, confirming the validity of the approach and identifying some of the challenges still ahead.

9.1 Introduction

The previous chapter focused on obstacle detection in environments that have varying amounts of texture. Successful obstacle detection may ensure successful obstacle avoidance, if the robot is able to stop (almost) immediately upon detection. This is approximately the case for slowly flying quadrotor MAVs. However, if the robot has to fly with a minimum forward speed, matters already become more difficult. The robot cannot just stop to hover, and has to take into account constraints such as the minimal speed and maximal turn rate. The minimal speed is an important factor for the difficulty level of collision avoidance, especially when regarded relatively to the density of obstacles in the environment [6]. Of course, the higher the obstacle density, the more difficult collision avoidance is.

Also the structure of the environment is important. Many collision avoidance studies focus on a forest-like environment only populated by trees as obstacles (e.g., [5–8]). After detection, such trees can typically be avoided by small deviations from the original flight trajectory. In addition, the avoidance of one tree hardly influences the probability of encountering a next tree. This is not so straightforward in typical indoor environments, where a wrong turn may bring the MAV to a corner from which an escape is no longer possible. This difficulty of indoor collision avoidance

© Springer Science+Bussiness Media Dordrecht 2016   167
G.C.H.E. de Croon et al., *The DelFly*, DOI 10.1007/978-94-017-9208-0_9

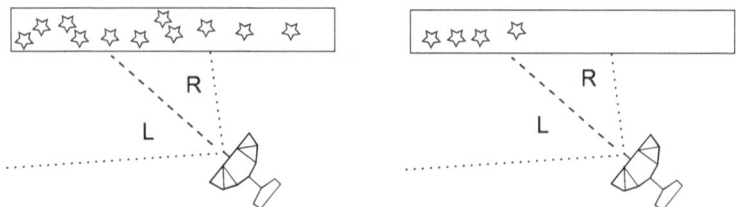

Fig. 9.1 *Left* an MAV approaches a wall with texture on both sides of the FoE. The flow can be used to determine the turning direction. *Right* an MAV approaches a wall that has some texture left of the FoE and little texture on the right. Straightforward comparison of the flow magnitude results in steering towards the wall

is aggravated by a small Field Of View (in the order of $\approx 60°$), which is typical for small cameras. Since all these factors are present for the case of the DelFly, it may be clear that collision avoidance is not an easy matter.

In the previous chapter, the control of the DelFly was reactive and straightforward: upon an obstacle detection it always turned right for a fixed time interval. Likely, this strategy is not ideal for avoiding obstacles on the long term. It may be advantageous to have at least a turning logic that determines when the MAV should turn left or right.

If the MAV is assumed to fly at a fixed height, the magnitude of the flow left and right from the Focus of Expansion (FoE) can be used for determining the direction in which the MAV should turn, as was done in [9]. The left part of Fig. 9.1 illustrates an MAV approaching a textured wall (stars represent textural features on the wall). If there is no rotational optical flow (or if inertial measurements can be used to derotate the optical flow), the flow in the right part of the image is stronger than that in the left part. The stronger flow indicates a shorter distance to the wall on the right, implying that the MAV should turn to the left.

The above-explained turning logic has been used in previous experiments with the DelFly II, but gave unsatisfying results. In cases where the obstacle contained little texture, as in the right part of Fig. 9.1, the MAV would take the wrong decision and steer towards the obstacle. Such cases happen often in indoor environments, especially with a small FOV. Obviously, the matter of texture-poor obstacles has to be taken into account in the turning logic.

In this chapter we extend the autonomous flight capabilities of the DelFly II, making **two contributions**. First, a novel turning logic is presented for deciding to turn left or right in situations where texture may be scarce. In addition, an onboard gyro is used to better stabilize the yaw-direction of the DelFly. Since this reduces the rotational component of the optical flow, it should lead to more reliable time-to-impact estimates [9]. Moreover, onboard control of the yaw-rate should allow for sharper turns. Second, the research presents the first work on the use of a pressure sensor onboard a flapping wing MAV for autonomous height control.

Below, we first explain the turning logic in Sect. 9.2.

9.2 Turning Logic

The main challenge for taking turning decisions is to deal with the uncertainty in the optical flow. This uncertainty derives from the sparsity of features and noise in the images. The turning logic introduced here explicitly takes into account the following factors:

1. Instead of comparing flow magnitude, the local time-to-impact estimates $\hat{\tau}$ are averaged on each side of the FoE. Consequently, $\tau_{RL} = \tau_R - \tau_L$ is the measure used for determining which way to turn: if the time-to-impact is larger on the right (τ_{RL} positive), the DelFly should turn to the right.
2. The number of features on the right and the left sides of the FoE (f_R, f_L) is an indication of the uncertainty of the τ_{RL} estimate. Fewer features on one of the sides more probably lead to erroneous estimates of τ_{RL}.
3. The suitability of the texture for determining optical flow and consequently time-to-impact can be automatically evaluated. The optical flow algorithm selects features at locations with high 'quality', i.e., where the minimal eigen value of the derivative covariation matrix is high with respect to neighboring locations. Averaging the minimal eigen value over the image is a good indication of the uncertainty in the optical flow estimates. This average is henceforth referred to as q.
4. A single estimate of τ_{RL} is likely not sufficient to determine a turning direction. Insufficient texture or all kinds of noise can render the estimate unreliable. Therefore, information should be integrated over time.

The factors above are incorporated in an algorithm that integrates information on τ_{RL} and its uncertainty over time. The DelFly Turning Algorithm (DTA) starts out with a turning decision $D_{RL} = 1$, with positive values representing turns to the right and negative values turns to the left. This decision is initially without evidence ($e = 0$). Subsequently, per image pair DTA first verifies that (i) the DelFly is not currently performing a turn (as measured by the median horizontal optical flow $u_x < T_{ux}$), (ii) the image pair contains features of sufficient quality ($q \geq T_q$), and that (iii) the number of features on each side is not too low ($\min(f_R, f_L) \geq T_f$). If on the basis of this verification, τ_{RL} is considered sufficiently reliable, the decision value D_{RL} is updated as follows:

$$D_{RL} \leftarrow \frac{(e D_{RL} + e' \tau_{RL})}{e + e'}, \tag{9.1}$$

where

$$e' = w_q q + w_f \min(f_L, f_R), \tag{9.2}$$

implying that the decision is a weighted average of τ_{RL} estimates. A pseudo-code representation of DTA is shown as Algorithm 1.

Algorithm 1 DelFly Turning Algorithm (DTA)

1: $D_{RL} \leftarrow 1, e \leftarrow 0$;
2: **while** flying **do**

3:     Determine optical flow and τ_{RL}
4:     **if** Commanding turn $\vee v_x < T_{vx}$ **then**

5:         $e \leftarrow 0$

6:     **else if** $q \geq T_q \wedge \min(f_R, f_L) \geq T_f$ **then**

7:         $e' \leftarrow w_q q + w_f \min(f_L, f_R)$
8:         $D_{RL} \leftarrow \frac{(D_{RL}e + \tau_{RL}e')}{e + e'}$
9:         $e \leftarrow e + e'$

10:    **end if**

11: **end while**

9.3 Overview of the Control Algorithm

There are two control loops that are separate from each other: the height control is performed by the microcontroller onboard the DelFly on the basis of the pressure sensor, while the flying direction is determined by a laptop that serves as 'ground station'.

For the obstacle avoidance experiments the height is controlled by the microcontroller onboard of the DelFly (an Atmega88PA). The SCP1000 barometer measures the pressure in the flight room at a frequency of 5 Hz to a resolution of 5 Pa which corresponds to about 40 cm. The pressure as measured at the start of the experiment is used as a reference value. PI-control with saturated integrator input and low pass filtered proportional term is employed for regulating the height, which can be done by increasing or decreasing thrust. As is well-known, the pressure can vary significantly over longer time spans, but currently such time spans are longer than the flight duration of the DelFly.

The ground station determines the flight direction by giving rudder commands to the microcontroller onboard the DelFly. These commands are: fly straight, turn to the left, and turn to the right. When a turn is commanded, the microcontroller sets a yaw reference value at $\pm 90°$. It then integrates the yaw rate measurements over time, actuating the rudder to perform the turn. The algorithm for deciding upon the rudder command to give is illustrated with a flow-chart in Fig. 9.2. In case of a collision detection, the turning logic is applied to determine the turning direction. Then the turn is commanded, while suppressing new turns for a predetermined 'turn

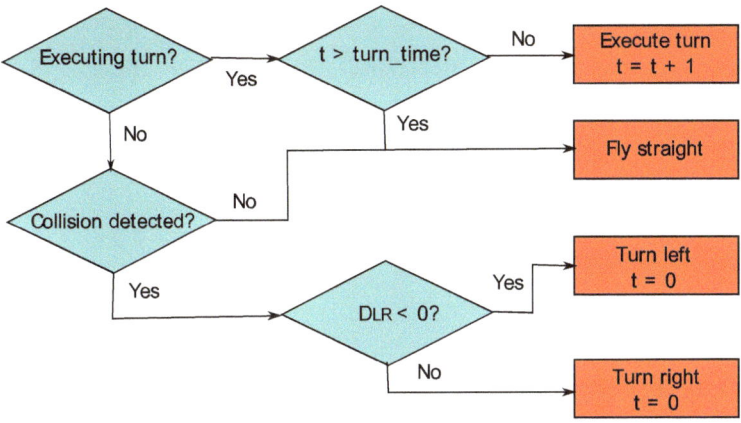

Fig. 9.2 Flowchart for the control of the rudder. See the text for details

time'.[1] The turning sequence is always executed fully before new commands are allowed. When not executing a turn, a collision is detected if: $(\tau_{OF} >= 0 \wedge \tau_{OF} < 2 \wedge \sigma_{OF} < 2 \wedge |\dot{\psi}| < 12°/\text{sec}) \vee \mathcal{H}(\hat{p}) < 2.2$. The values for the thresholds have been determined in simulation and subsequently tuned on the real platform [3]. The condition on $|\dot{\psi}|$ is used for preventing new detections during a turn. In order to prevent the DelFly from getting into a downward spiral, the yaw rate is limited with the help of the onboard gyro. When flying straight, the onboard microcontroller runs a P-controller that attempts to keep the yaw rate at 0. The onboard running P-controller results in rather straight flight trajectories in between turns.

The video frame rate is 30 Hz. The frames are down-sized to 160×120 pixel images and processed by both the optical flow algorithm running at 30 Hz and the texton method running at 13 Hz. The execution frequencies have been measured on a dual-core 2.26 GHz laptop with all ground station software running at the same time.

9.4 Experiments

9.4.1 Simulation

In [4], we briefly mentioned simulation experiments that were performed in order to explore and compare different strategies and parameter settings. In particular, we stated that using the DelFly Turning Algorithm led to a much longer flight time than

---

[1]Since there is no communication yet from the microcontroller to the ground station, the ground station uses a fixed time instead of a signal that indicates turn completion.

Table 9.1 Performance of different turning algorithms in simulation. See the text for details

Method	Average flight time ($\pm\sigma$)	Percentage no collision (%)
Right turning, 90°	65.0 ($\pm$ 37.3)	0
Immediate τ_{RL}, 90°	60.6 ($\pm$ 44.4)	0
DelFly Turning Algorithm, 90°	**87.8** ($\pm$ 51.0)	0
Right turning, 180°	140.5 ($\pm$ 90.7)	10
Immediate, τ_{RL} 180°	114.2 ($\pm$ 75.6)	0
DelFly Turning Algorithm, 180°	**171.2** ($\pm$ 98.1)	**25**

just basing the turning direction on the instantaneous τ_{RL}. Here, we reproduce and extend the simulation experiments. As a first extension, we compare the left-right turning controllers with a pure right turning controller. As a second extension, we look at the effect of the turning angle by performing tests with a turn angle of $\sim$90° and of $\sim$180°.

Simulation experiments took place in the same simulation environment as described in Chap. 8, with a texture surface proportion of 0.20. The simulated DelFly was initialized in the middle of the room with a random initial direction. The DelFly's height was fixed to 1 m, while the flight direction was determined by the rudder control in Fig. 9.2. Rudder commands are perfectly executed, i.e., they result in a constant yaw rate. The DelFly is subject to small 'drafts' (in the order of 0.1 m/s) that can slightly change the velocity and heading. The simulation ended either when the DelFly collided with a wall in the rectangular room, or when it reached 300 s without any collision.

Table 9.1 contains the results of three methods: (1) Right turning, (2) Immediate τ_{RL}, and (3) DelFly Turning Algorithm. Results are mentioned for turns of 90° and 180°. The table shows the average flight time ($\pm$ standard deviation σ), and the proportion of runs without any collision. Per condition, $n = 20$ simulation runs were performed.

The results in Table 9.1 confirm the earlier finding that DTA considerably improves upon the use of an immediate measure of τ_{RL}. For a turning angle of 90°, the average increase in flight time is 27.2 s, while for 180° it is 57.0 s.

The inclusion of a right turning strategy is interesting. Generally, it performs better than the instantaneous τ_{RL} strategy and worse than the DTA. Figure 9.3 shows a failure case of just right turning for a turn angle of 180°. This type of situation is handled better by DTA. However, for a turning angle of 180° in an empty rectangular room this situation does not seem to arise so often. In addition, making a left-right decision also represents a risk of taking the wrong decision. This may explain the bad performance of using the instantaneous τ_{RL} for left-right turning.

This leads us to perhaps the most surprising outcome of the experiments: the turn angle has a much larger influence on flight time than the difference in turning algorithm. Our hypothesis is that in the empty rectangular room used in simulation, a turn angle of 180° simply leads to fewer encounters with walls. Also, it likely influences

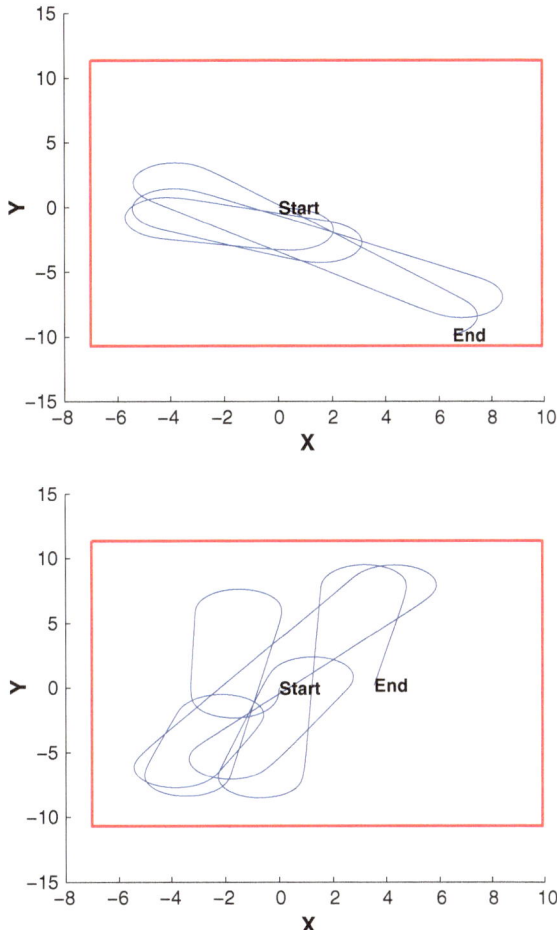

Fig. 9.3 *Top* Trajectory of simulated DelFly that only turns to the right 180°. *Bottom* Trajectory of simulated DelFly that turns right or left 180° based on the DelFly Turning Algorithm

how often a situation occurs in which the MAV has to change turning direction. In order to get more insight in this matter, a more elaborate study would be required in which also other parameters are varied and tested in different environments (with different shapes and obstacles inside the environment).

Finally, please note that the standard deviation of the flight times is quite large for all conditions. We used the double-tailed bootstrap method [2] to test whether the flight times of the different methods were drawn from different distributions. However, this could not be shown with statistical significance for any combination of conditions. This may change if more simulation runs are performed and the maximal flight time is augmented—DTA 180° reached the maximal flight time in 25 % of the cases.

9.4.2 Real World

The real-world obstacle avoidance experiments performed for this study have as goal to test the turning logic in the presence of texture-poor obstacles. For the real tests, the parameters used in simulation were further tuned for the real DelFly. The resulting parameter settings are: $T_{vx} = 48$ pixels, $T_q = 10^{-4}$, $T_f = 10$, $w_q = 5000$, and $w_f = 0.1$.

We analyze the data from an experiment that illustrates both the strengths and limits of the DTA discussed in Sect. 9.2—selecting a test that ends in a collision. Videos of this and other experiments can be found online.[2] The (x, y)-positions of the DelFly during the flight are shown in the top part of Fig. 9.4, where the cross marks the start of autonomous control and the circle the end of the trial. The left, top, and right lines of the box illustrate a wall with a couch, a wall with a whiteboard, and windows with closed blinds, respectively. The bottom line represents the table with the ground station. The bottom part of Fig. 9.4 shows the height of the DelFly over time.

From the experiments, we make four main observations. The first observation concerns the height control with the pressure sensor: in spite of the low-frequency and low-accuracy measurements, the PI-controller successfully controls the DelFly's height within safe margins. The variation in the height is ± 40 cm around the set point. The average height over the experiment does not vary over the time span of this experiment. In addition, sudden changes in pressure (due for example to the opening of doors) are successfully filtered out. Importantly, the local pressure variations due to the flapping of DelFly are not significant anymore after application of the PI controller. The height controller can be switched on before flight and still functions correctly when in hover flight.

The second observation is a confirmation of the findings from the literature (e.g., [9]) that the turning direction tends to remain the same over time in a rectangular room. Often the DelFly turns right as this is the initial value given to D_{RL}. Indeed, if we take a look at the trajectory in Fig. 9.4, all turns until the last one are to the right.

The third observation is that DTA's integration of information over time and use of the reliability of time-to-impact estimates is absolutely necessary to take correct turning decisions. Figure 9.5 shows all the signals related to the turning logic, all scaled with constant factors for visualization purposes: D_{RL} (black line, scaled by 0.2), τ_{RL} (red, 0.25), τ_R (dark blue, dashed, 0.25), τ_L (dark blue, dotted, 0.25), the quality q of the features (orange dashed, 10^4), the rudder command (purple, 1), and the heading flow u_x (light blue, 3). Time steps at which the signals allow for an update of D_{RL} are indicated with a cross. Below the signals, the onboard images are shown.

The most important signal is τ_{RL}, the difference in time-to-impact between the right and left side of the FoE. Its value is often wrong. Around 362 s it is negative as a consequence of little texture and around 365.8 s it is negative due to the execution

---

[2]http://www.youtube.com/watch?v=_p6a8ei4PZc.

Fig. 9.4 *Top* (x, y)-position of the DelFly II over time during one of the experiments. The top, left, and right lines of the box illustrate the walls in the room, the bottom line the table with the ground station. *Bottom* height of the DelFly during the test

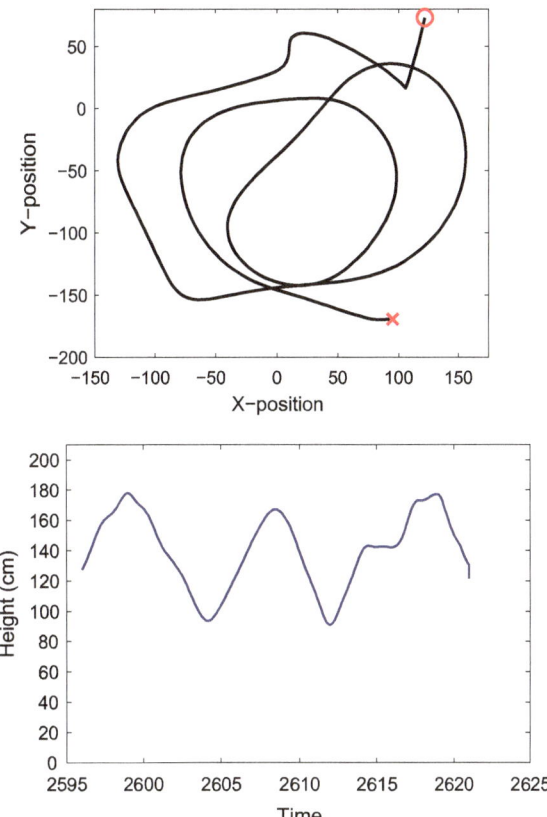

of a turn. These wrong values are respectively detected by means of $q < T_q$ (little texture), and a large heading flow (plus active rudder command). As a consequence of ignoring the negative values, the DelFly does not turn left when detecting an obstacle at 365.9 s.

The fourth observation is that with the current speed and the confined space in which the experiment is performed, the DelFly spends a large time executing turns. While executing turns, the evidence e is reset and no new evidence is gathered. As a consequence, it often happens that the DelFly bases its turning decision on few measurements. A few measurements that are mistakenly considered 'reliable' can have a detrimental effect on the turning decision D_{RL}. Figure 9.6 shows how noise in the image leads to a wrong estimate of τ_{RL} just after a turn. Because there is an obstacle detection, the DelFly turns to the left immediately, bringing it into a corner from which it cannot turn away in a timely fashion.

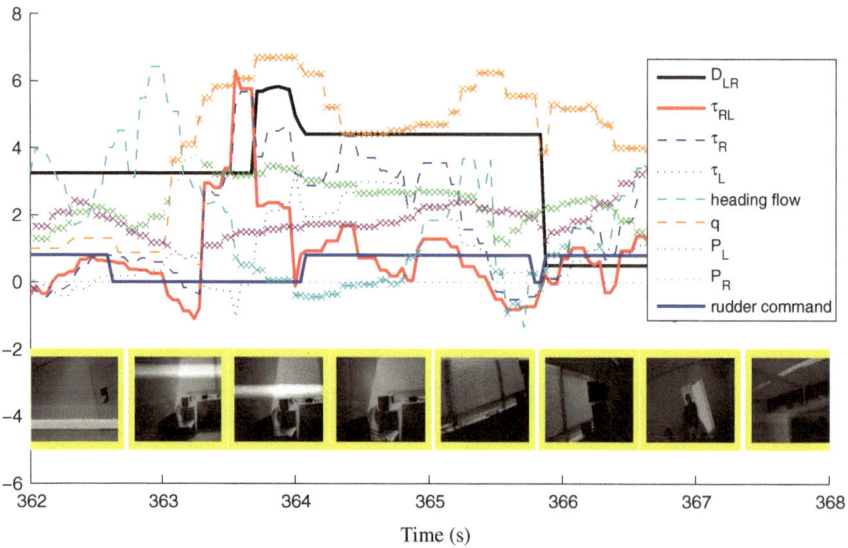

Fig. 9.5 All signals relevant to the turning logic (best viewed in color, see text for details)

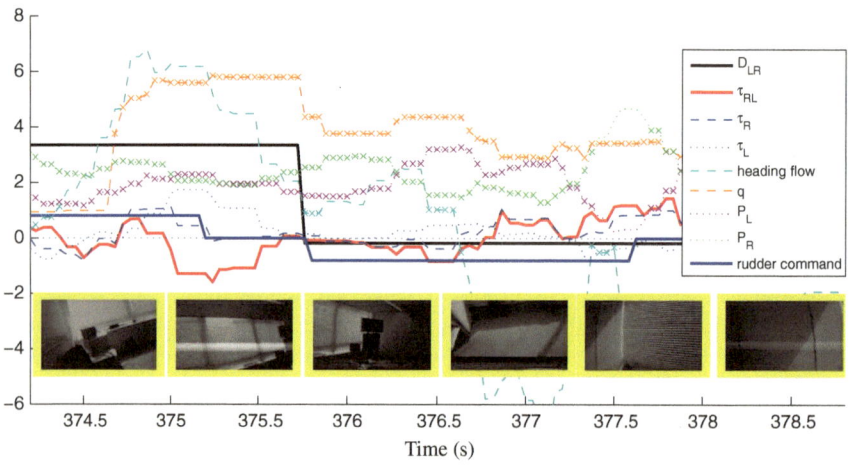

Fig. 9.6 All signals relevant to the turning logic (best viewed in color, see text for details)

9.5 Conclusion

With regards to the DelFly's autonomous flight capabilities, a step forward was made both in height control and obstacle avoidance. The height control was successfully performed by the onboard microcontroller using a pressure sensor. The pressure differences due to the flapping flight did not have a detrimental effect on height control. Still, the height variations were in the range of ± 40 cm, which means that there is room for improvement. Additional information from accelerometers or vision may help to reduce the height variations and may provide a means to keep a safe height also for longer time spans. Concerning the obstacle avoidance, a step forward was made by introducing an improved turning logic. Together with onboard control of the yaw rate, this led to improved obstacle avoidance capabilities with respect to instantaneous use of the flow magnitude. Still, faulty decisions occur, resulting in collisions with indoor obstacles. One small improvement of the DTA may be to not reset the evidence e to 0, reducing the influence of only a few τ_{RL} measurements. As main directions for larger improvements of the obstacle avoidance capabilities, we identify the incorporation of onboard vision processing for reducing noise and allowing the direct feedback of gyro measurements (cf. [9]) or the back-EMF of the motors (as in [1]) for derotating optical flow. Neuromorphic vision sensors with a higher update frequency, sensitivity, and a larger field-of-view may lead to further improvements of obstacle avoidance capabilities.

Acknowledgments This chapter is partly based on [4].

References

1. F.G. Bermudez, R. Fearing, Optical flow on a flapping wing robot, in *IROS 2009* (2009), pp. 5027–5032
2. P. Cohen, *Empirical Methods for Artificial Intelligence* (MIT Press, Cambridge, 1995)
3. G.C.H.E. de Croon, E. de Weerdt, C. de Wagter, B.D.W. Remes, R. Ruijsink, The appearance variation cue for obstacle avoidance. IEEE Trans. Robot. **28**(2), 529–534 (2012)
4. G.C.H.E. de Croon, M.A. Groen, C. de Wagter, B.D.W. Remes, R. Ruijsink, B.W. van Oudheusden, Design, aerodynamics, and autonomy of the delfly. Bioinspir. Biomime. **7**(2) (2012)
5. D. Dey, K.S. Shankar, S. Zeng, R. Mehta, M. Talha Agcayazi, C. Eriksen, S. Daftry, M. Hebert, J. Andrew (Drew) Bagnell, Vision and learning for deliberative monocular cluttered flight, in *Field and Service Robotics (FSR)* (2015)
6. S. Karaman, E. Frazzoli, High-speed flight in an ergodic forest, in *2012 IEEE International Conference on Robotics and Automation (ICRA)*. IEEE (2012), pp. 2899–2906
7. L. Matthies, R. Brockers, Y. Kuwata, S. Weiss, Stereo vision-based obstacle avoidance for micro air vehicles using disparity space, in *2014 IEEE International Conference on Robotics and Automation (ICRA)*. IEEE (2014), pp. 3242–3249
8. S. Ross, N. Melik-Barkhudarov, K.S. Shankar, A. Wendel, D. Dey, J.A. Bagnell, M. Hebert, Learning monocular reactive uav control in cluttered natural environments, in *IEEE International Conference on Robotics and Automation (ICRA)* (2013)
9. J.-C. Zufferey, *Bio-inspired Flying Robots: Experimental Synthesis of Autonomous Indoor Flyers* (EPFL/CRC Press, Lausanne, 2008)

Autonomous Flight with Onboard Stereo Vision

10

Abstract

We shift our focus to the use of stereo vision for autonomous flight. Stereo vision implies carrying two cameras on board, which adds weight and increases the power consumption. Still, it also allows for instantaneous distance estimates, which is a considerable advantage on a moving (and oscillating) flapping wing MAV. In particular, we explain the onboard stereo vision and control algorithms that allow the 20-g DelFly Explorer to autonomously fly around in unknown environments for as long as its battery lasts.

10.1 Introduction

In Chap. 8 it was observed that the use of optical flow on the DelFly is hampered by its motion and flapping oscillations. The camera images are distorted due to the line-by-line recording, invalidating the use of a linear camera model. The core problem here is the use of a standard CMOS camera on a flapping wing MAV. Hence, a possible solution direction is to change the camera sensor to have specifications closer to those of insect facet eyes. Specifically, eyes of for instance bees have a lower spatial resolution and much higher temporal resolution [1]. There is a large effort in robotics to make such optical sensors, even mimicking the optic flow processing that happens just behind the facet eyes [2–5].

Another solution direction is to determine distances *instantaneously*, with the help of stereo vision. In stereo vision, two cameras are placed next to each other at a known fixed base distance. Due to their different vantage point, the left and right cameras can observe the same object at a different horizontal position in the image. Objects close by will have a large horizontal displacement between the images, while objects far away will have the same horizontal position. With knowledge of the camera's properties the displacement in pixels (called *disparity*) can be used to determine the distance in meters to the given world point. The key idea here is that if the individual (horizontal) image lines are synchronized between the left and right camera, the vertical distortions caused by the flapping motion are much less relevant. In addition, the disparities are directly related to distances in the environment, while optical flow can only convey distances relative to the (unknown) velocity of the DelFly.

Despite the above-mentioned advantages of stereo vision, there are also some disadvantages. First, carrying a second camera implies more mass and energy consumption of the vision system. Second, algorithms for stereo vision processing are known to be computationally intensive, with real-time processing typically reserved for dedicated hardware architectures like Field Programmable Gate Arrays (FPGAs). Third, even if we succeed in estimating the distances to points in the environment, it is not clear what algorithm can provide robust obstacle avoidance for a flapping wing MAV. The DelFly is not yet able to hover all by itself, so a computationally efficient algorithm has to be devised that takes this system constraint into account. Fourth, making a small-scale stereo vision system with synchronized image lines is a challenge in itself.

The fourth challenge of making the electronic hardware was already discussed in Chap. 4, where we have shown and discussed two stereo vision systems that were created in the course of our stereo vision research for DelFly autonomy. Initially, a system was made consisting of two cameras and an image transmitter. Later, a system was made with digital cameras and an STM32F4 processor able to process the images on board.

In the remainder of this chapter, we discuss our approach to the first three challenges, and how this approach allows the *DelFly Explorer* to perform fully autonomous flights, avoiding obstacles for as long as its battery lasts. The structure of the chapter is as follows. First, we present the DelFly Explorer platform, and its hardware modifications that allow its autonomous flight (Sect. 10.2). Subsequently, we explain the obstacle avoidance algorithm, which is specifically tuned to the DelFly's size and flight characteristics, and which ensures obstacle-free flight under a few reasonable conditions (Sect. 10.3). Then we introduce the stereo vision algorithm that runs on board of the DelFly Explorer, investigating both its accuracy and computational effort (Sect. 10.4). This novel strategy is tested in unmodified real-world environments Sect. 10.5. We conclude in Sect. 10.6.

10.2 DelFly Explorer

The *DelFly Explorer* is an illustration of how an all-round approach to the design of flapping wing MAVs can lead to increased autonomy capabilities. The advances made in aerodynamic understanding, design, and electronic systems result in a platform able to carry a 0.98-g autopilot (accelerometers, gyros, magnetometers, barometer) and a 4.0 g stereo vision system (cameras and processor). This (rather heavy) payload allows the autonomous exploration of unknown spaces, hence the new name. The DelFly Explorer has a wing span of 28 cm and a weight of 20 g.

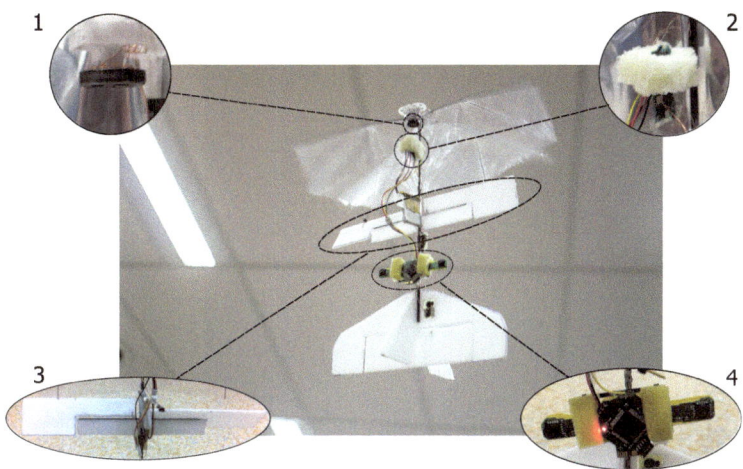

Fig. 10.1 Picture of the *DelFly Explorer*. The *four insets* show the main changes with respect to the DelFly II: **a** the number of windings in the brushless motors has been reduced to cope with the Explorer's higher weight, **b** an autopilot with a complete IMU, barometer, and an ATmega328P—MLF28 microcrontroller, **c** the DelFly explorer uses ailerons behind the wings instead of a rudder on the tail, and **d** the onboard stereo vision system with STM32F405 processor for onboard vision processing

The DelFly Explorer is shown in Fig. 10.1, with insets showing its four main innovative components. The first inset shows the brushless motor. The number of windings around the coils has been reduced from 37 to 32. This way the ratio of RPM (and hence lift) versus input voltage is increased at the cost of a lower torque. As a result the lift generated at 3.5 V is still sufficient to keep the heavier DelFly Explorer in the air. This is in contrast to the old case where it would descend when the voltage dropped below 3.9 V. The flight time of the DelFly Explorer is typically around 8–10 minutes.

The second inset shows a side-view of the autopilot, including an ATmega328P—MLF28 microcontroller, 3-axis accelerometers, gyros, magnetometers, and a barometer. Furthermore, it features two-way telemetry and RPM-monitoring. The autopilot is not necessary to achieve stable flight, as the tail of the DelFly passively stabilizes it during flight. However, the autopilot can serve other purposes, such as performing height control, disturbance rejection or more precise attitude control.

The third inset shows a set of ailerons placed just behind the wings. These ailerons are necessary for making smooth turns, which is essential to autonomous flight. The DelFly II featured a rudder for making turns. Deflection of the rudder first caused the DelFly II to yaw (around the Z-axis—see Fig. 10.2 for the axes definition), which in turn also resulted in a heading change. However, the yaw rotations during turns rendered computer vision processing during turns problematic. The ailerons of the DelFly Explorer make the DelFly roll (around the X-axis), and since it flies close to up-right, this directly influences the heading without creating any rotations of the camera images.

Fig. 10.2 Sketch of DelFly
Explorer in flight with the
body-axes definition

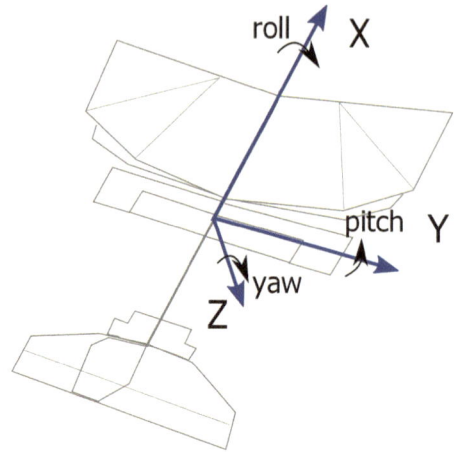

Finally, the fourth inset shows the stereo vision system in more detail. As explained in Chap. 4, it has two digital cameras with a baseline of 6.0 cm and an STM32F405 processor. Importantly, the flapping motion of FWMAVs introduces deformations in the camera images [6,7]. Therefore, it is not possible to use subsequently recorded left and right images for stereo matching [8]. The cameras of the stereo system are synchronized and provide $YUYV$ image streams, and in the current implementation a Complex Programmable Logic Device (CPLD) merges the streams from both cameras by alternately taking the Y component of the stream from both cameras. This results in a single image stream with the order $Y_l Y_r Y_l Y_r$ (where l stands for left and r for right). The resulting stream contains simultaneously sampled pixels at full camera resolution but without color.

10.3 Obstacle Avoidance Algorithm

10.3.1 Overview of Obstacle Avoidance Strategies

The DelFly Explorer uses the disparity maps to avoid obstacles in its environment. Obstacle avoidance is a long-standing and major topic in robotics (see, e.g., [9] for an overview). Of course, the difficulty of obstacle avoidance is determined by the properties of the robot and its environment. As mentioned, if the robot moves slowly and can stop at any moment, obstacle avoidance is easier and mostly determined by its obstacle detection capabilities. Obstacle avoidance is also easier if the environment has only a few obstacles.

Concerning the difficulty of the obstacle avoidance task, it is important to note that the DelFly Explorer in its current form cannot yet hover and is a nonholonomic vehicle, akin to robot cars [10,11] or fixed wing UAVs [12,13]. Unfortunately, most existing obstacle avoidance methods on such vehicles are not appropriate for the case of the DelFly. First, most methods assume a perfect knowledge on the positions of the robot and obstacles. For instance, in [13] 3D-navigation of a fixed wing UAV is studied, in which an artificial potential field is generated by Dipolar Navigation Functions. However, all obstacles are known beforehand, also those beyond the current field-of-view. Second, many methods involving path planning or optimization [10,11,14–16] are computationally too demanding for the small processor onboard the DelFly Explorer. A major reason for this is that they do not just aim to avoid obstacles, but also try to minimize the deviation from a predetermined path toward a goal state [10].

Given the computational constraints on board the DelFly Explorer, we focus on computationally extremely efficient approaches. For instance, in [14,15], a planned path of a fixed wing UAV is augmented with a reactive obstacle avoidance method that places triangles around detected obstacle points. If multiple triangles overlap, it is assumed that the UAV has to fly around the exterior of the group of triangles. Although in many outdoor environments with sparse obstacles this method will provide good results, it is unsuited for the closed indoor environments in which the DelFly Explorer will have to navigate. There are even more efficient approaches to obstacle avoidance. A notable example of this type of approach is [17], in which a 10-g indoor flyer avoids the walls in a small room by turning left when the optic flow on its right is higher and vice versa. This approach can be called truly 'reactive', in the sense that the controller is a function that maps the current percept of the environment directly to an action [18]. In our previous work on monocular obstacle avoidance with the DelFly [19] we have used a similar approach to [17]. However, we extended the algorithm with a bit of memory in order to prevent 'chattering' (turning left/right/left/etc.), making our approach not reactive in the strict sense of the definition given above.

The above-mentioned *left/right strategy* can easily run into problems in indoor environments, even given a perfect obstacle detection. The left part of Fig. 10.3 shows

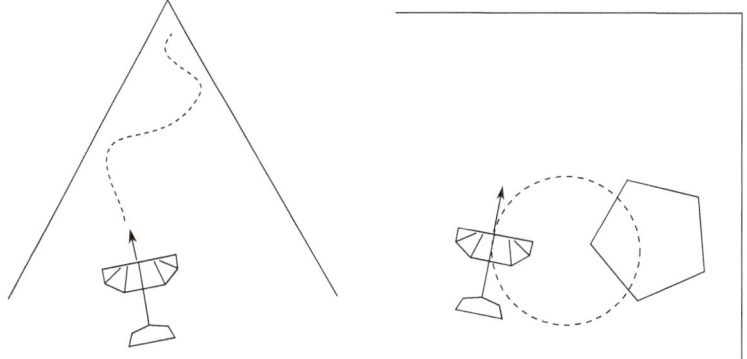

Fig. 10.3 Failure cases of a left/right turning strategy. *Left* The robot turns away from the closest obstacle but gets caught in a funnel. *Right* The robot turns away from the obstacle but runs into an obstacle outside its field-of-view

a prototypical failure case. It is a funnel that will inevitably trap the MAV, as is shown by the dotted trajectory. The right part of Fig. 10.3 shows another prototypical failure case. As in the example, there can be obstacles in the line of the full-turn circle, which are currently outside of the field-of-view. When they arrive in the field-of-view, it may be impossible to avoid a collision.

10.3.2   Proposed Obstacle Avoidance Strategy

In [20], we proposed a strategy that ensures obstacle avoidance, taking into account the constraints of a nonholonomic vehicle with a limited field of view. This *droplet* strategy is illustrated in Fig. 10.4. The central idea is to always keep an empty area in the field of view that is large enough to make a full turn. This leads to a 'droplet' shaped area, of which the size and shape are determined by the minimum turn radius r, the margin m, and the angle of the field-of-view α. As soon as the robot detects an obstacle that threatens to enter the droplet area, a fixed maneuver is performed in which it first flies with an angle β to the principal axis of the stereo cameras, and then makes a full turn in front of the obstacle. To achieve the angle β, the stereo vision system is mounted with this angle to the DelFly's body. It is normally 14°, so Fig. 10.4 exaggerates the magnitude of the angle for illustration purposes. When turning, as soon as the robot sees a new direction that fits at least an entire droplet, it can go straight again. In contrast to the previously mentioned approaches, the proposed strategy ensures an obstacle-free robot trajectory. That is, a theoretical

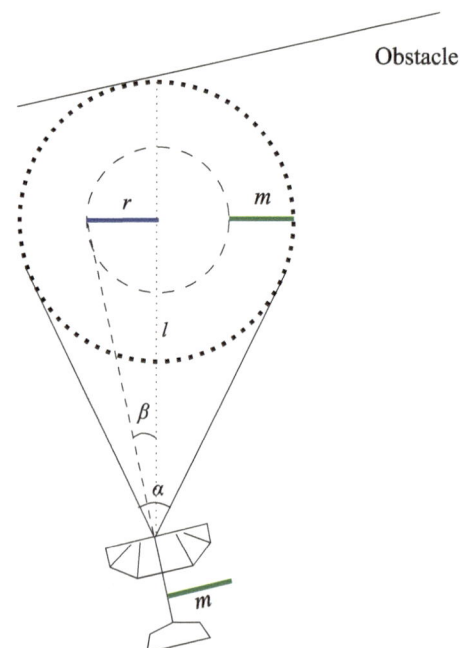

Fig. 10.4 Illustration of the droplet strategy. The central concept is to keep an open space that allows a full turn maneuver within the field of view. Hence, the droplet shape is determined by the field of view α, the turn radius r, and the margin m. The length of the droplet l depends on these variables and for the current hardware is 3 m. After detection, the DelFly flies at an angle β from the principal axis of the stereo cameras, until it starts turning. This means that the stereo cameras are mounted with an angle β with respect to the DelFly's body

point-sized robot with perfect distance sensing will encounter no obstacles on its path.

To take into account the robot's body, the strategy incorporates the margin m. It covers half the body length plus a safety margin, for instance to account for external disturbances or distance estimation inaccuracy during the maneuver. If given a random initial position, a robot with body risks colliding during its trajectory with angle β toward the full turn. The reason for this is that at the start of the maneuver, its body will surpass the border of the initial field-of-view. However, in a behavioral context this situation will not occur, since such an obstacle would have entered the droplet area at a previous time instant.

The biggest difference between the theory behind the strategy and its real world application lies in the imperfect sensing and actuation. On the sensing side, the stereo vision of the DelFly Explorer is too noisy to accurately determine small obstacles entering the droplet area. We use quite crude measures to determine obstacle presence: for instance, the number of pixels in the field-of-view that are closer than the droplet depth. To account for noise, a threshold on this number of pixels is set to trigger the avoidance maneuver. Obviously, this will lead to crashes with small obstacles that remain too long below the detection threshold. On the actuation side, the DelFly Explorer will not make the perfect droplet maneuver, as it can be perturbed by external influences such as strong drafts. There is also a variation in the turn radius, if it is not closed loop controlled. Actuation inaccuracies are especially dangerous if the DelFly continues its flight too long before making the full turn. An adequate margin m prevents most such troubles though.

The droplet strategy is implemented as a small Finite State Machine (FSM). We illustrate the FSM in Fig. 10.5. The normal behavior (state 1) is to fly straight. If an

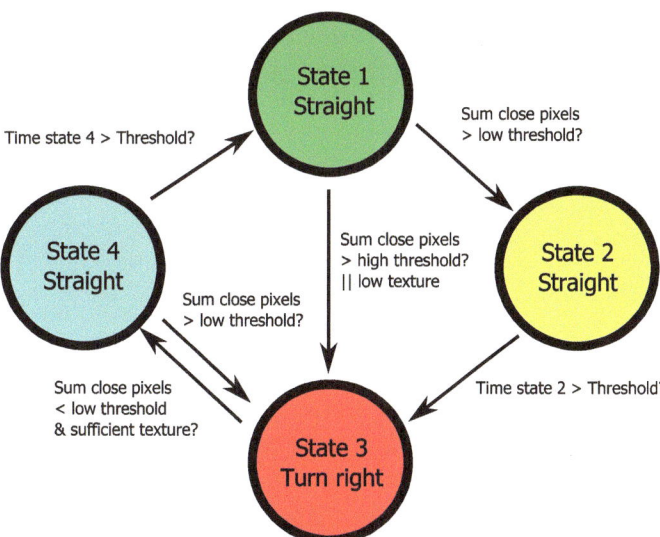

Fig. 10.5 Implementation of the droplet strategy with a finite state machine

obstacle is detected at the border of the droplet region (the sum of close pixels is larger than a low threshold), the DelFly will fly the fixed maneuver. This maneuver first requires the DelFly to still fly straight (state 2) and after some time start circling (state 3). If the DelFly discovers a direction in which the obstacles are far enough away, it will start flying straight again. It will first do so carefully (state 4), so that it can immediately start turning again on the first sign of obstacles (transition to state 3). Normally, after some time it will transition to state 1 again. There is an additional transition from state 1 to state 3 in the case an obstacle is suddenly close to the DelFly or in the case there is very little visual texture.

10.4 Stereo Vision

For the stereo vision system carried on board the DelFly, a new stereo vision algorithm was developed that is presented in this section. For autonomous obstacle avoidance, the DelFly requires real-time processing of the stereo images in combination with sufficient depth quality. Since the stereo system is heavily restricted in terms of processing speed (168 MHz) and memory availability (max. 192 kB RAM), it is important to find the right point on the trade-off between speed and quality.

Among the huge number of stereo vision algorithms that can be found in literature there are two groups that are not regarded to be suitable for our application. These are the algorithms that perform global optimization, and the algorithms that are based on local matching. The first group is too demanding in terms of power and memory requirements, while the second group provides insufficient quality when dealing with image regions that contain little texture. In between these groups there is another group of algorithms that perform semi-global optimization. Examples of these algorithms are 1-D Dynamic Programming [21] and Semi-Global Matching [22]. These algorithms perform optimization along certain individual directions. The drawback of such an algorithm is that an error somewhere along this optimization line has an effect on the rest of the optimization line. These effects are limited in [22] by optimizing over multiple directions. However, this increases the required amount of processing and memory again.

10.4.1 LongSeq

For these reasons a new algorithm is proposed that performs optimization along one image line at a time, where badly matched pixels do not have a degrading effect on the matching quality. For reasons to become clear in the explanation, we call it the *LongSeq* algorithm. The first step in the algorithm is to compute the matching costs $C(x, d)$ of the pixels in one image line by calculating the absolute difference in intensity for a disparity range d_{range} starting from a minimal disparity d_{min}.

$$C(x, d) = |I_l(x) - I_r(x - d)| \qquad (10.1)$$

Then the minimum matching cost $C_{min}(x)$ for each pixel is computed:

$$C_{min}(x) = \min_d C(x, d) \tag{10.2}$$

Based on these cost measures (matching cost and minimum matching cost), a binary image B is computed for all pixels and disparities of the image line using two thresholds: τ_{cost} and τ_{min}:

$$B(x, d) = \begin{cases} 1 & \text{if } C(x, d) > \tau_{cost} \text{ and } C_{min}(x) < \tau_{min} \\ 0 & \text{otherwise} \end{cases} \tag{10.3}$$

The cost threshold τ_{cost} is used to define if a pixel match is good or bad. A matching cost above the threshold indicates a bad match. The minimum cost threshold τ_{min} is used to check if there is at least one disparity value for which the pixel has a good match. $B(x, d)$ will only be nonzero when pixel x has a some good matching candidate, but if that is not the case for the disparity value considered. Pixels that have no good matching candidates are simply ignored. As a result, image B indicates which pixels have better candidates at other disparities. All other pixels are ignored at this stage since they have either no good matching candidate, or they match well at the considered disparity value.

The next step is to find sequences of neighboring pixels in an image line that do not have better matching candidates at other disparities (i.e. $B(x, d) = 0$). The length of this sequence will be used as a measure for matching quality and it is therefore stored in image B. This is done by replacing all zero values by the length of the sequences they belong to. For example, let us consider eight neighboring pixels (50–57) in a line for one disparity value, e.g., 7. From Eq. 10.3 the following fictitious values were obtained:

$$B([50\ 57], 7) = [1\ 0\ 0\ 0\ 1\ 0\ 0\ 1]$$

This series of values contains two sequences of zeros; one with length 3 and one with length 2. The zeros in B then are accordingly replaced by these numbers.

$$B([50\ 57], 7) = [1\ 3\ 3\ 3\ 1\ 2\ 2\ 1]$$

An initial disparity map D_{init}^{left} is then computed by selecting from B for all x the disparity value with the highest number (longest sequence):

$$D_{init}^{left}(x) = \max_d B(x, d) \tag{10.4}$$

The matching cost as described in Eq. 10.1 is defined for matching the left image with the right image. The process is repeated for matching the right image with the left image to obtain D_{init}^{left} and D_{init}^{right}. These disparity maps can now be combined to optimize the result. This is done by mapping the left disparity image to the right disparity image:

$$D_{map}^{left->right}(x - D_{init}^{left}(x)) \leftarrow -D_{init}^{left}(x) \tag{10.5}$$

The optimal disparity is then found by taking the minimum of the two disparity maps:

$$D_{opt}(x) = \min(D_{map}^{left->right}(x), D_{init}^{right}) \tag{10.6}$$

This optimization step is required to handle disparity discontinuities. The algorithm is named *LongSeq*, because it favors long sequences with constant disparity in an image line. In situations where there is little to no texture, this will slightly bias the result to high-disparity estimates. In the context of obstacle avoidance, this is very sensible: low-texture images often occur close to obstacles and in any case present a danger, since they do not provide information on distances to obstacles ahead.

This method assumes that the images contain only fronto-parallel planes. Furthermore it specifically tries to match image planes with low variation in texture. By sliding these planes over each other, there will be one disparity where the overlap between the planes from the left and right image will reach its maximum. This effect is measured by the length of the sequences, and for this reason the maximum length is selected as the best match.

The proposed method shows some similarities with plane sweeping algorithms [23] in that it tries to match an image plane for a certain orientation. However, in the proposed method only fronto-parallel planes are considered for computational reasons. Moreover, in contrast to [23], LongSeq searches the largest line section meeting this assumption.

The stereo vision system on board the DelFly Explorer does not yet have any wireless connection for sending images during flight. Therefore, we show results of the stereo system in-hand, with the images sent via a serial connection. Figure 10.6 shows nine examples of stereo vision images and their corresponding disparity maps. The left column shows the left images, the center column the right images, and the right column shows the disparity maps, in the interval [0, 10] (bad pixels are also set to 0). Please remark that even though the camera is held in hand, the images already have motion deformations and blur. The examples show the types of obstacles we expect that the DelFly should detect. The top six examples show results for medium-sized obstacles such as poles while the bottom three examples show obstacles that cover large areas such as walls.

The line-wise matching strategy of the proposed algorithm can be clearly seen in the images by the striping effects. By observing the detected poles it can be noted that texture poor areas tend to have the same disparity as the poles. This effect might be reduced in some cases by using more complex algorithms that perform optimization in more directions. Note that this effect is less of an issue, since it will sooner lead to an unnecessary obstacle detection than a missed detection. In general the background will be assigned the same disparity as the pole and not the other way around. Exceptions occur in situations where the contrast between the pole and the background is very low. The first six examples in Fig. 10.6 show that the presence of the poles is clearly indicated.

The results from example 7 and 8 in Fig. 10.6 are far from perfect, but the results are useful for our application. Example 7 shows that the wall is fairly close to our camera, even though the structure in the middle is the only feature that provides sufficient texture. In the case of example 8, the algorithm is able to indicate that the bottom part of the images contain obstacles at at smaller range compared to the rest of the image.

Fig. 10.6 Nine examples of the stereo vision processing. The columns show from *left* to *right*: the left image, the right image, and the disparity image produced by the proposed stereo vision algorithm LongSeq. The disparity images are color coded from low-disparity (*dark*) to high-disparity (*bright*)

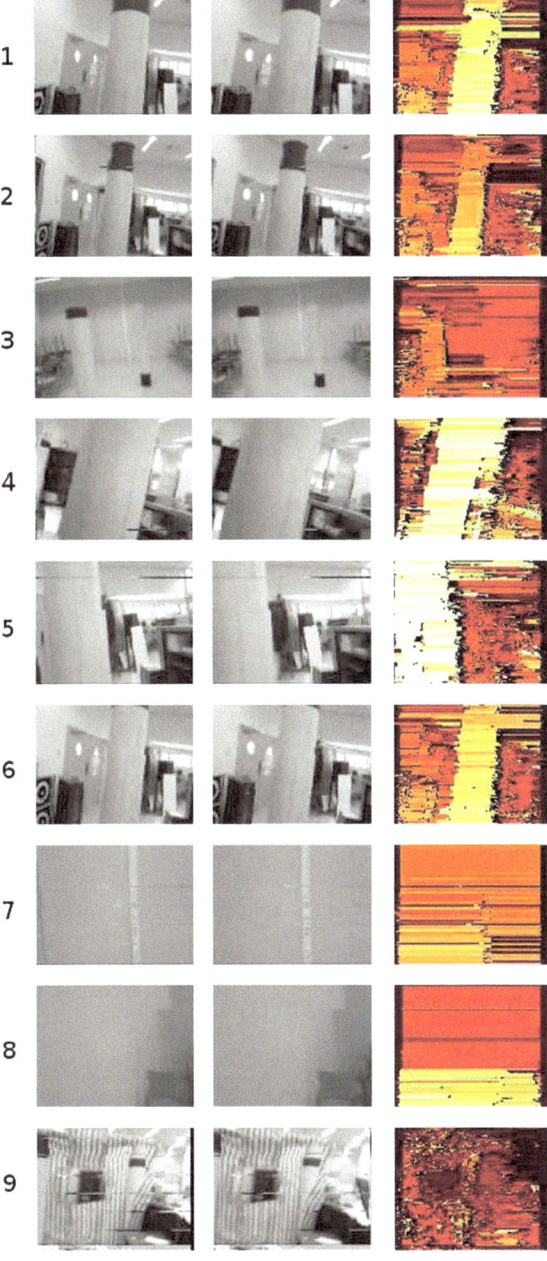

The effect of the pixel-based matching cost is illustrated by example 9 of Fig. 10.6. The dense variation in contrast in combination with the low resolution images results in many small sequences and a large variation of disparity values. This effect might be reduced by using windows for calculating the matching cost but this increases computational load as well as memory requirements.

10.4.2 Subsampling

In the interest of computational efficiency, typical stereo vision steps such as undistortion and image rectification are skipped. Without these steps, LongSeq takes around 90 ms of processing on the STM32F405 on a full image of 128 × 96 pixels. Hence, it runs at ~11 Hz. For many applications of the stereo vision system 11 Hz can be sufficient. However, for some applications, such as obstacle avoidance or flying through a window, a higher processing frequency may be desired. The same goes if one wants to perform additional vision tasks besides stereo vision.

If the interest is not in dense 3D scene reconstruction, but some type of aggregate disparity values are used (as in [8,24]), then *sub-sampling* can be applied. Sub-sampling typically leads to a considerable gain in computational efficiency at a low cost in accuracy [25]. Since LongSeq is line-based, a natural way of sub-sampling is to process fewer lines.

The control algorithm explained in Section 10.3.2 bases its decisions on the number of pixels in the image with a disparity higher than 5. This implies that the detailed disparity maps are aggregated into only two values. Hence, it makes sense to apply sub-sampling for achieving higher processing frequencies. Figure 10.7 shows the number of processed image lines versus the processing times as measured on the

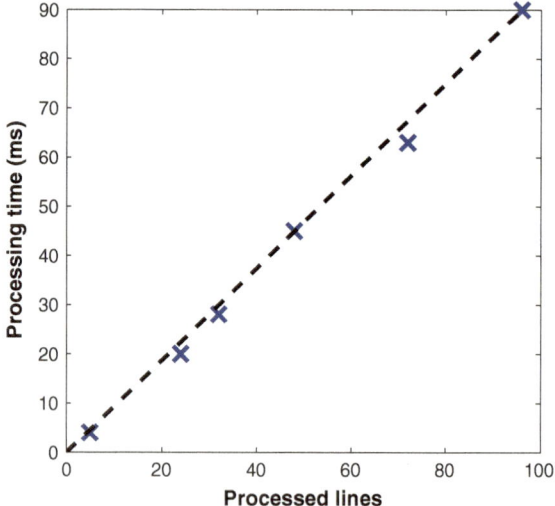

Fig. 10.7 Processed image lines versus processing time onboard the DelFly

Fig. 10.8 Effect of subsampling on the aggregate value used by the obstacle avoidance control algorithm, the total number of pixels with a disparity higher than 5 (close pixels). The results are shown for various subsampling ratios, ranging from 5 % (*red*) to 100 % (*green*). The lines of the extreme percentages (5 and 100 %) are drawn *thick*

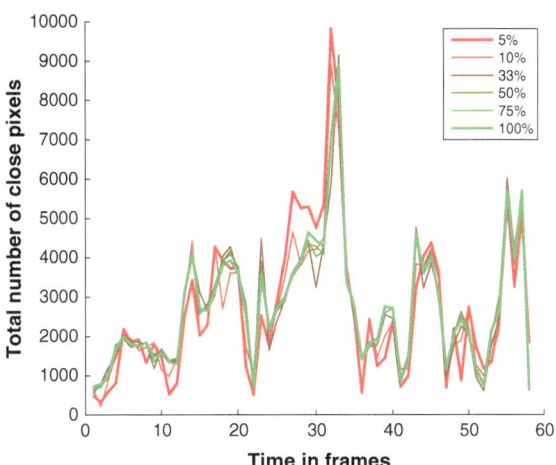

STM32F405 (crosses). As to be expected, this relation is roughly linear (dashed line). In order to process at frame rate, one can sample 32 image lines (one third of the image).

Figure 10.8 shows the effects of sub-sampling on the estimated number of pixels with a disparity larger than 5 (top) and on the difference between the left and the right image (bottom). As can be seen in the figure, all sampling ratios follow the trend of the case of full sampling (100 %)—albeit with a variation that increases with a decreasing sampling ratio. Surprisingly, this is even valid for a low sampling ratio of 5 % (4 image lines out of 96 in our implementation).

10.5 Real-World Experiments

The droplet strategy has been implemented both on a DelFly II carrying the stereo vision system with offboard processing (Fig. 4.8) and on the DelFly Explorer carrying the 4-g stereo vision system with onboard processing (Fig. 4.10).

Three tests were performed with the DelFly II with offboard stereo vision processing. The tests took place in the MAV-lab at Delft University of Technology (see Fig. 10.9). The flights were captured by two external GoPro cameras, so that the trajectory of the DelFly could be reconstructed. During the tests, heading control was performed autonomously, while height control was done by a human operator. The DelFly carried two batteries: one for flying and one specifically for the vision system. The tests lasted as long as the batteries allowed.

The reconstructed trajectories can be seen in Fig. 10.10. The most important observation from Fig. 10.10 is that no crashes occurred. In test 2 and 3 the trajectory's red line intersects with an obstacle's thick black line on the right, but this is because the DelFly was flying higher than the obstacle. In fact, each test ended, because the

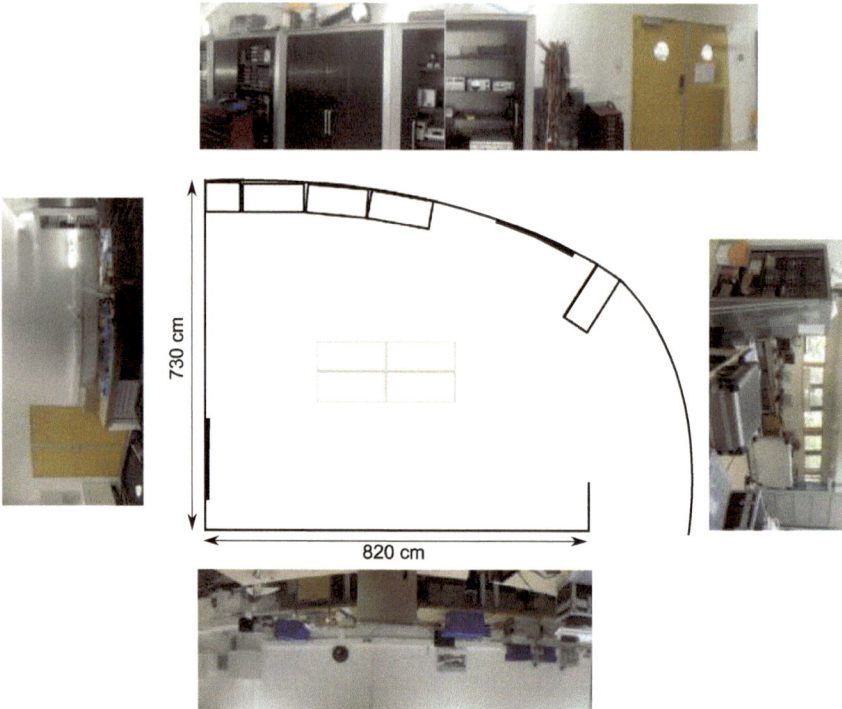

Fig. 10.9 Floor plan of the MAV-lab, where the experiments took place, together with panorama images illustrating the appearance of the room

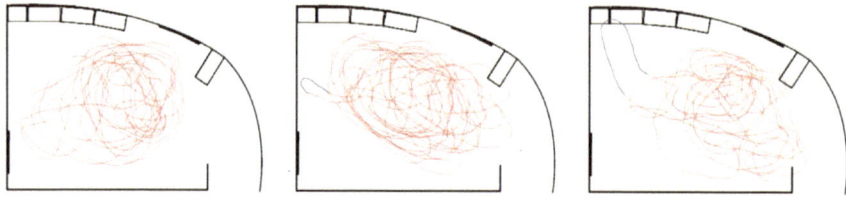

Fig. 10.10 Flight trajectories of the DelFly during three different experiments. The second and third experiment both show a part of the trajectory in *blue*, indicating that this part was manually added because of the vehicle flying out of the camera field of view

battery of the stereo vision system drained. As soon as the operator noticed this, he flew the DelFly back to the starting point. Another observation to make is that by following the simple droplet behavior the DelFly explores a large part of the lab. However, there are some parts of the lab that are visited less often, such as on the bottom left. This area has little texture, which may be the reason it is typically avoided.

Table 10.1 Quantitative results of three flight tests

Test	1	2	3
Total time	4 m 55 s	6 m 21 s	4 m 40 s
Total distance	151.3 m	185.6 m	99.5 m
Average speed	0.51 m/s	0.49 m/s	0.35 m/s
Number of turns	38	41	32
Fraction of time turning	0.17	0.22	0.27

Fig. 10.11 Time lapse image of a part of an experiment in a lecture room—reprinted with permission. For the complete video, see https://www.youtube.com/watch?v=2KPgPWb2Gkg

Quantitative results of the three tests are shown in Table 10.1. It shows the total time, total distance, average speed, number of turns, and fraction of time spent turning. A few observations are in place here. First, the average speed differs over the different tests. This is due to a manual trimming of the elevator at the start of the run, which leads to variations in pitch angle and hence speed. Second, Table 10.1 shows that the DelFly was turning for 17–27 % during the flights. Still, Fig. 10.10 shows that most parts of the trajectories are curved. This is due to an offset in trim of the aileron, and variation is also partly caused by battery placement and small misalignment. Third, the total flight times and distances also show the effectiveness of the droplet avoidance strategy. As mentioned above, the flight time was mainly limited by the duration of the battery powering the stereo vision system.

Similar experiments have been performed with the DelFly Explorer, carrying a single battery that powered all systems. It also performed multiple autonomous collision-free flights for as long as its battery lasts (8–10 min). A few of the videos can be found on YouTube.[1] Figure 10.11 shows a time lapse of part of an experiment

---

[1] https://www.youtube.com/watch?v=2KPgPWb2Gkg.

in one of TU Delft's lecture rooms. Please note the red light on the stereo board on one of the DelFly's images: it indicates that the stereo vision system detected an obstacle, in this case the wall behind the camera capturing the trajectory.

10.6 Conclusion

In this chapter, we presented our approach to achieve autonomous collision-free flight with the help of stereo vision processing on board of a Flapping Wing MAV. The 20-g DelFly Explorer can perform fully autonomous flights with the help of the 'droplet' strategy, which is implemented on the onboard 4-g stereo vision system.

Unfortunately, this does not mean that the collision-avoidance problem is solved. Although the DelFly Explorer is the first light-weight FWMAV able to fly in unknown environments, it cannot yet reliably fly in *any* unknown environment. A part of this is explained by the current hardware, since very thin obstacles such as electricity cables will not even be picked up by the stereo vision algorithm, and definitely not by the control. There are other obstacles that may be hard to detect, such as large, fully transparent windows. If one wants to detect these windows, it may be necessary to increase the level of vision reasoning. However, potentially a design can be found so that the DelFly can easily recover from the occasional collision with a window (as small bugs typically also do not detect the window but bump into it until they get tired with it). Advances in hardware, vision processing, and control are necessary to cover the remaining obstacles which are not successfully avoided.

Finally, collision avoidance cannot be the final goal of the DelFly. Concerning exploration, there is also still a long way to go. For instance, the current control strategy focuses on safe flight, which is to some extent detrimental for exploration. A solution to this may lie in further improvements of the system, allowing the DelFly to adopt a much smaller droplet region. However, randomly flying around with the droplet strategy is likely not the most efficient way to explore an unknown environment. Hence, more goal-directed behavior should be developed that takes into account what the DelFly has already seen, where there are passage-ways to other rooms, etc.

Acknowledgments This chapter is partly based on [8, 26]. We would like to thank Sjoerd Tijmons, who has been a major contributor to the work on stereo vision.

References

1. M.F. Land, D.-E. Nilsson, *Animal Eyes* (Oxford University Press, New York, 2002)
2. D. Floreano, R. Pericet-Camara, S. Viollet, F. Ruffier, A. Brückner, R. Leitel, W. Buss, M. Menouni, F. Expert, R. Juston, M.K. Dobrzynski, G. LEplattenier, F. Recktenwald, H.A. Mallot,

N. Franceschini, Miniature curved artificial compound eyes. *Proc. Natl. Acad. Sci. U.S.A, PNAS* **110**(23), 9267–9272 (2013)

3. N. Franceschini, J.M. Pichon, C. Blanes, J.M. Brady, From insect vision to robot vision. Philos. Trans. Biol. Sci. **337**(1281), 283–294 (1992)
4. F. Ruffier, N.H. Franceschini, Aerial robot piloted in steep relief by optic flow sensors, in *(IROS 2008)* (2008), pp. 1266–1273
5. G. Sabiron, P. Chavent, T. Raharijaona, P. Fabiani, F. Ruffier, Low-speed optic-flow sensor onboard an unmanned helicopter flying outside over fields, in *IEEE International Conference on Robotics and Automation* (2013)
6. F. Garcia Bermudez, R. Fearing, Optical flow on a flapping wing robot, in *IROS 2009* (2009), pp. 5027–5032
7. G.C.H.E. de Croon, E. de Weerdt, C. de Wagter, B.D.W. Remes, R. Ruijsink, The appearance variation cue for obstacle avoidance. IEEE Trans. Robot. **28**(2), 529–534 (2012)
8. S. Tijmons, G.C.H.E. de Croon, B.D.W. Remes, C. De Wagter, R. Ruijsink, E-J. Van Kampen, Q. Chu, Stereo vision based obstacle avoidance on flapping wing mavs, in *EuroGNC* (2013)
9. F. Bonin-Font, A. Ortiz, G. Oliver, Visual navigation for mobile robots: a survey. J. Intell. Robot. Syst. **53**(3), 263–296 (2008)
10. A. Kelly, B. Nagy, Reactive nonholonomic trajectory generation via parametric optimal control. Int. J. Robot. Res. **22**(7–8), 583–601 (2003)
11. E. Papadopoulos, I. Poulakakis, I. Papadimitriou, On path planning and obstacle avoidance for nonholonomic platforms with manipulators: a polynomial approach. Int. J. Robot. Res. **21**(4), 367–383 (2002)
12. M. Hwangbo, J. Kuffner, T. Kanade, Efficient two-phase 3d motion planning for small fixed-wing UAVs, in *IEEE International Conference on Robotics and Automation (ICRA)* (2007), pp. 1035–1041
13. G. Roussos, D.V. Dimarogonas, K.J. Kyriakopoulos, 3d navigation and collision avoidance for nonholonomic aircraftlike vehicles. Int. J. Adapt. Control Signal Process. **24**(10), 900–920 (2010)
14. S. Griffiths, J. Saunders, A. Curtis, B. Barber, T. McLain, R. Beard, Maximizing miniature aerial vehicles. IEEE Robot. Autom. Mag. **3**, 34–43 (2006)
15. S. Griffiths, J. Saunders, A. Curtis, B. Barber, T. McLain, R. Beard, Obstacle and terrain avoidance for miniature aerial vehicles, in *In Advances in Unmanned Aerial Vehicles* (Springer, Netherlands, 2007), pp. 213–244
16. D. Jia, J. Vagners, Parallel evolutionary algorithms for uav path planning, in *Proceedings of the AIAA 1st Intelligent Systems Conference* (2004)
17. J.-C. Zufferey, A. Klaptocz, A. Beyeler, J.-D. Nicoud, D. Floreano, A 10-gram microflyer for vision-based indoor navigation, in *Proceedings of the IEEE/RSJ International Conference on Intelligent Robots and Systems (IROS)* (2006)
18. S. Nolfi, Power and the limits of reactive agents. Neurocomputing **42**(1–4), 119–145 (2002)
19. G.C.H.E. de Croon, M.A. Groen, C. De Wagter, B.D.W. Remes, R. Ruijsink, B.W. van Oudheusden, Design, aerodynamics, and autonomy of the delfly. *Bioinspiration Biomimetics* **7**(2), 025003 (2012)
20. S. Tijmons, G.C.H.E. de Croon, B.D.W. Remes, C. De Wagter, R. Ruijsink, E. van Kampen, Q.P. Chu, Off-board processing of stereo vision images for obstacle avoidance on a flapping wing mav, in *Pegasus AIAA conference, Prague* (2013)
21. S. Forstmann, Y. Kanou, J. Ohya, S. Thuering, A.Schmitt, Real-time stereo by using dynamic programming, in *Computer Vision and Pattern Recognition Workshop (CVPRW)* (2004), pp. 29–29
22. H. Hirschmuller, Accurate and efficient stereo processing by semi-global matching and mutual information. Comput. Vision Pattern Recogn. **2**, 807–814 (2005)
23. D. Gallup, J.M. Frahm, P. Mordohai, Q. Yang, M. Pollefeys, Real-time plane-sweeping stereo with multiple sweeping directions, in *Computer Vision and Pattern Recognition (CVPR)* (2007), pp. 1–8

24. S. Tijmons, G.C.H.E. de Croon, B.D.W. Remes, C. De Wagter, R. Ruijsink, Obstacle avoidance by stereo vision on flapping wing mavs (Submitted)
25. G.C.H.E. de Croon, C. De Wagter, B.D.W. Remes, R. Ruijsink, Sub-sampling: real-time vision for micro air vehicles. *Robot. Auton. Syst.* **60**(2), 167–181 (2012)
26. C. De Wagter, S. Tijmons, B.D.W. Remes, G.C.H.E. de Croon, Autonomous flight of a 20-gram flapping wing mav with a 4-gram onboard stereo vision system, in *2014 IEEE International Conference on Robotics and Automation (ICRA 2014)* 2014

Part IV
Conclusions

Conclusions and Future Research

11

Abstract

In this chapter we reflect upon the questions that have been answered and, as importantly, raised in the course of the DelFly project. We identify some of the core problems remaining in the field and discuss the ongoing and planned future work on the road to an ever smaller and more autonomous DelFlys.

11.1 Conclusions

Looking back on the DelFly project up until the publication of this book, we mainly discern as a red line the influence of the real world on the problems we studied and the scientific approach we adopted. Our main motivation is to engineer smaller, more capable flapping wing MAVs (more robust, enduring, autonomous, etc.) for observation missions in the real world. Tests with the DelFly in "normal" environments, such as the autonomous obstacle avoidance tests in our coffee corner,[1] inevitably show the inadequacy of the current design and algorithms—often leading to new scientific problems that are to be solved.

11.2 Future

As hinted to in the previous section, there are still many open questions. For instance, we do not yet have simulations allowing the automatic design of flapping wing MAVs. For a given design it is unclear a priori what transient or cycle-averaged forces and moments will be generated during operation.

---

[1] See https://www.youtube.com/watch?v=5Cm0CvLdbSQ.

© Springer Science+Bussiness Media Dordrecht 2016
G.C.H.E. de Croon et al., *The DelFly*, DOI 10.1007/978-94-017-9208-0_11

While in the past we have always focused on the design, aerodynamics, and artificial intelligence of the DelFly, in the future we foresee at least one additional main research area: system identification. This makes for a total of four research areas, which we will discuss in turn in the remainder of this section.

11.2.1 Design

Future designs of the DelFly may be influenced by yet unknown advances. For instance, significant advances in battery or actuation technology may have a big influence on the design. These are possible advances though that are outside of our own reach.

We currently focus on the following main design choices. First, we work toward DelFlys that are more maneuverable and can better deal with windy conditions. This implies striving for a higher wing loading (a higher mass/wing surface ratio). If the DelFly gets smaller, while retaining the same weight, the DelFly will be less influenced by wind gusts. In order to counteract wind gusts, a better maneuverability is necessary, and hence taking actions with a higher force and at higher speed. This may be difficult to achieve with actuators on the tail, and argues for direct actuation with the wings. Of course, direct actuation with the wings would lead to a significantly different design as well.

Another path we are currently exploring, is more integrated incorporation of more advanced electronics on board. After the introduction of the autopilot, we are now aiming for incorporation of electronics in the structural elements of the DelFly. 3D-printing can also play its role in such a more integrated electronics design.

11.2.2 Aerodynamics

An important part of the DelFly II design process is the understanding and further optimization of the related flapping-wing aerodynamic mechanisms. In this book, we presented some of the preliminary studies that were carried out to gain better understanding of the aerodynamic force generation by clap-and-fling motion and to identify the effects of wing geometry and flexibility on the force production so that a better aerodynamic performance can be attained by use of an improved wing design. However, such an optimization study is not straightforward due to the complicated nature of flexible flapping wing aerodynamics. This complexity is even further enhanced by the wing-wing interaction that is employed as a force enhancement mechanism in the DelFly flapping flight. The lack of analytical solutions and empirical models for this particular multi-parameter problem makes it necessary either to simplify the subject by focusing on a certain aspect of the mechanism (e.g., use of rigid wings to study the generic flapping-wing aerodynamics or use of flexible wings with isotropic structural properties) or to use advanced experimental and numerical tools to investigate the mechanism in its complete form in real flight conditions including all interrelated aerodynamic, inertial and structural phenomena. The first strategy has been followed in many studies in the literature and it is very useful to gain the fundamental understanding of the effects of individual aspects (e.g., motion typology, kinematics, wing geometry, wing structural properties, etc.) on the flapping-wing aerodynamics. Yet, the outcome of these studies cannot be superposed to explain their combined effects due to nonlinear interactions occurring between the aforementioned phenomena. On the other hand, improvement in the capability of flow measurement systems [11] and increasing computational power in the last decade offer a great potential in studying the flapping-wing flight of MAVs and biological flyers in its complete form.

In this respect, one of the future works for the DelFly aerodynamics team is to develop a three-dimensional numerical flow model of the DelFly flapping-wing flight, which takes into account the complex fluid-structure interaction that occurs between the unsteady flow fields and flexible wings of anisotropic structural character. This tool will be an important step forward in the design and optimization of the DelFly II to achieve a better aerodynamic performance. As a first step, numerical flow models and meshing strategies have been developed and tested without structural deformation calculations. Instead these models make use of prescribed wing deformations, which are acquired in time-resolved stereo-vision measurements.

Another future goal is to perform three-dimensional flow field measurements during a free-flight of the DelFly II in a wind tunnel. In all previous measurements, the experimental DelFly II model was mounted on a balance mechanism, which limits the dynamic oscillations of the ornithopter that are otherwise present in free-flight conditions. Therefore, the flow field measurements in free-flight conditions can reveal some important flow dynamics that were not captured in the previous studies. Besides, such measurements can serve as a comparison and validation basis for numerical investigations.

11.2.3 System Identification

One of our current focus areas is the 'system identification' (or modelling) of the DelFly II in flight. High-end motion tracking systems allow the tracking of several points of the DelFly's body during free flight. Post-processing of these points over time allows to determine the positions, velocities, forces and moments during different flight maneuvers. It is already interesting to compare these forces to the ones measured when the DelFly is clamped to a measurement balance [6,7]. However, the final goal of system identification is to create a model that predicts future states based on the current state and the control actions. A highly accurate model can be used for gaining insight in the (control) dynamics of flapping wing MAVs. However, such a model that possibly will predict time-varying effects during the flapping cycle, will likely be computationally expensive. A less accurate, but computationally more efficient model could be invaluable for onboard control of the flapping wing's state.

In preliminary work, we have investigated different types of linear, cycle-averaged models. First, a 'white-box' model was employed, in which we fitted parameters of the model to bouts of the flight trajectories [5]. Although the measured forces and moments were predicted rather accurately, this did not suffice for simulating the

DelFly, which resulted in divergence. Subsequently, a linear, time-invariant 'black-box' model has been investigated [1]. This implies fitting the parameters of the system matrices to the flight trajectories. The measurements again matched quite accurately, and this time also simulation was possible. Unfortunately, the black box parameters do not give insight into the physical processes underlying the behavior. An investigation into the poles and zeros of the fitted linear systems did show good correspondences between different fits.

In future work, we will focus more and more on models that are valid for larger parts of the flight envelope, give more insight into the actual physical phenomena at work, and that also explain time-varying effects.

11.2.4 Artificial Intelligence

From the start of the DelFly project, it was foreseen that the onboard camera could be used for autonomous flight. The first tests involved following a person with a red suitcase (by means of color detection). Over the years, the DelFly platform in general and the camera in particular have improved continuously, up to the point that we now have a platform with onboard vision processing. These improvements have allowed for a gradual increase in the autonomous flight capabilities of the DelFly, going from paper trail following at the EMAV 2008 to fully autonomous obstacle avoidance in 2013. While in the past we have used terms as 'vision-based navigation' [8] and 'autonomous flight capabilities' [9], we are now at a point that we start using the term 'artificial intelligence' (as testified by the title of this book).

In order to illustrate this point, please think of the straightforward Finite State Machines (FSMs) we have used throughout the project for control. Already when designing such FSMs for competitions, we noticed that the complexity of these controllers increases if one takes into account more and more failure cases. The same goes for an increase of task complexity ('First take-off, stay within a circle of 1 m diameter for at least 10 s, then pass through a window, cross the obstacle zone, …'—from the indoor competition at the IMAV 2013). Making robust controllers for more complex tasks is difficult and requires a lot of trial-and-error by trying different controllers out in simulation.

In order to create robust controllers for more complex tasks, it would be a good idea to have an artificial evolutionary process design the controllers for us. However, a major difficulty with evolutionary robotics is that it is not easy to cross the 'reality gap' between the simulated and the real world. Typically, neural controllers are evolved in simulation. When the evolved neural controllers are implemented on the real robotic platform, they prove inadequate for the task due to the many properties of the robot and real world that have not been modelled in simulation. Typically, the controllers have to be retrained on the real robot in the real world in order to get successful results. With the DelFly, such a setup is impractical.

As a partial solution to the reality gap, we have recently introduced the use of *behavior trees* for evolutionary robotics [12] (see Fig. 11.1). Behavior trees are controllers that have their origin in the AI of computer games. In comparison to FSMs,

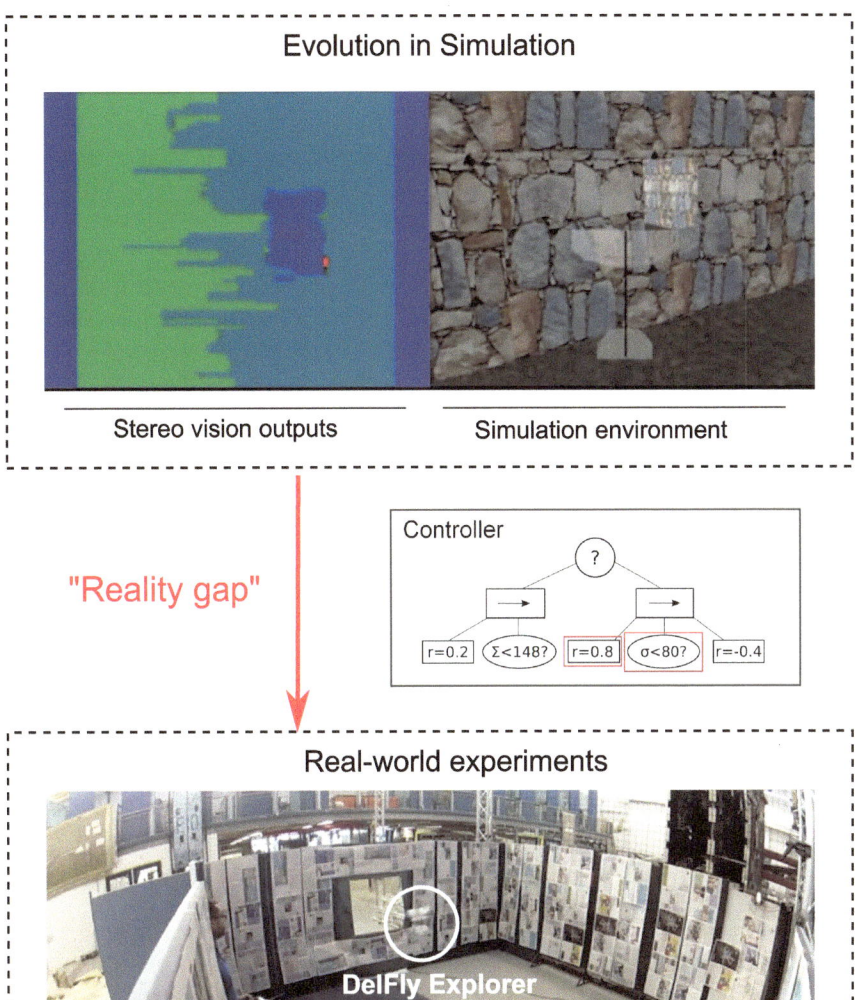

Fig. 11.1 Evolutionary Robotics setup in which the controller is evolved in simulation (*top*) and then transferred to the real robot (*bottom*). The behavior tree controller evolved in [12] is shown in the *middle*. The parameters in the tree that were modified manually for use on the real robot are highlighted in *red*. For a video on the experiments, see https://www.youtube.com/watch?v= CBJOJO2tHf4

they do not have the same problem of a complexity explosion. In comparison with neural networks, they are far easier to understand for the human designer. This last property allows the human to re-tune some of the evolved parameters easily in a few experiments with the real robot and thus helps to reduce the reality gap.

In our experiments, we evolved a behavior tree for the DelFly with which it could fly around in a room, search for a window, and pass through it. The evolved behavior tree was compared to a baseline behavior tree that was designed by ourselves. The evolved tree was not only smaller, but also had a slightly superior performance. In simulation, it passed through the window 88 % of the time versus 82 % of the time of the human-designed tree. Straightforward application of the controllers in the real world led to complete failure, but because of the comprehensibility of the behavior trees, retuning was straightforward. After retuning, the real DelFly Explorer passed through the window in our experimental environment 54 and 46 % of the times, using the evolved and human-designed tree respectively. The bottom part of Fig. 11.1 shows a run with the evolved behavior tree in which the DelFly passes through the window. All failure cases of the evolved tree were due to collisions with the window frame, likely caused by the draft around the window.

These first results are of course far from perfect. However, they lead to believe that better results are around the corner if we combine the comprehensible evolved behavior trees with somewhat better modeling of the DelFly and its environment. The main goal is to maintain computationally efficient controllers, while incrementally increasing the task complexity. For a single DelFly this will signify a step from pure obstacle avoidance to exploration, and from exploration to navigation (e.g., first explore a building and then return to a home position). When the autonomous capabilities of a single DelFly have reached a certain level, it will also become useful to have many DelFlys performing a task together, as a 'swarm'.

11.3 Application of Flapping Wing MAVs

Our main interest in this book has been in the underlying scientific questions and techniques relevant for designing autonomous light-weight flapping wing MAVs. Although different goals have been mentioned, such as the performance of an "observation mission", no concrete applications have been named. In this last section of the book we would like to indicate what we think are the unique properties of flapping wing MAVs as compared to other types of MAVs, and what applications flapping wing MAVs could be used for.

11.3.1 Properties of Flapping Wing MAVs

As has become apparent in this book, the design of flapping wing MAVs is much less developed than for instance the design of rotary or fixed wing MAVs. The flapping mechanism and the unknown unsteady aerodynamic effects introduce extra complexity in the design, and the flapping movements may complicate autonomous flight of flapping wing MAVs. These design challenges raise the question: are there reasons for which one should prefer a flapping wing MAV over other types of MAVs?

One major possible reason would be aerodynamic efficiency—at small scales, flapping wings become more and more efficient compared to fixed and rotary wing concepts. This does not mean that any small flapping wing is more efficient than any small rotorcraft. For instance, it was recently shown that the *Black Hornet* MAV developed by ProxDynamics is as efficient in hover flight as the flapping wings of the hummingbird [10]. However, the hummingbird can more easily change flight regime and fly efficiently in forward flight: it is able to migrate over large distances without the intake of new energy. This advantage of flying animals is in general not matched yet by flapping wing MAVs, but the flight duration of for instance the 16-gram DelFly II of 22.5 min in forward flight is promising in this respect.

Other reasons for preferring a flapping wing MAV are perhaps more of a practical nature and have become apparent to us during the many tests, demonstrations, and competitions we performed with the DelFly. Perhaps the most important is the speed at which the wings move during the flapping cycle. In general, the wings move much slower compared to rotary wings. This influences how bad it is to accidentally hit an object or person. Both a light-weight flapping wing MAV and rotary wing MAV are inherently very safe. However, the rotors of the small rotary wing MAV have to turn at high speeds to generate enough lift. The rotor may inflict a little bit of damage or pain when it comes in direct contact with an obstacle or person. In addition, hitting any object with the rotor will typically result in a crash and hence an end-of-mission. To remedy this, a hull (as, e.g., on the Parrot AR drone [4]) or a cage-like structure [2,3] is necessary. A flapping wing has already a lower speed of its wings. In addition, the speed is 0 at the outer ends of the flapping cycle. This means that typically the speed is lowest when it hits something, avoiding damage or pain. In addition, it also means that a collision does not always result in a crash.[2] In indoor environments, this property of coping with collisions is invaluable, as it will be extremely difficult to ensure a capability of a 100 % obstacle avoidance (think of how insects often fly into windows).

We would like to end with a reason that is usually not explicitly stated and perhaps not entirely objective: In our experience, and in contrast to rotary wings or fixed wings, people love to see a flying DelFly. Perhaps it reminds them of beautiful flying animals such as butterflies. As a consequence, people do not mind if the DelFly flies close to them.

Because of their safety and beauty, we think that flapping wing MAVs will be very suitable for indoor environments also populated by humans.

---

[2]For a few occasions in which we have live imagery of this situation, please look at https://www.youtube.com/watch?v=4Mxq-nr9xyY (at 2:10 in the video), or, even clearer https://www.youtube.com/watch?v=lOec0p8pNpA (starting from 1:28 in the video).

11.3.2 Example Applications

The previous section made clear what we think are the advantages of flapping wing MAVs. In this brief subsection, we mention some example applications.

1. **A swarm of DelFlys in greenhouses**: the swarm could monitor the health and ripeness of the crops in the greenhouse, deposit small worms that function as anti-parasites, or pollinate flowers.
2. **Emergency response**: a swarm of DelFlys may be set out after an earth quake to help locating survivors in partially collapsed or burned out buildings.
3. **A swarm of DelFlys finding a gas leak**: as a parallel to fruitflies finding rotten fruit, a swarm of DelFlys could locate a gas leak at an industrial site. The swarm could quickly cover the entire terrain, with DelFlys clustering around likely gas source locations.
4. **DelFlys as amusement park fairies**: with some modifications in appearance, a DelFly could look like a real flying fairy and could possibly also interact with children by means of a small microphone/speaker.

In short, we think that flapping wing MAVs can be functional, friendly, and fun. Of course, the list above is far from exhaustive and may not even contain an application that will actually be realized in the future. Still, by writing this book, we hope that more people will be able to contribute to the technology of flapping wing MAVs and to further unravel the mysteries of flapping wing flight.

References

1. S.F. Armanini, C.C. de Visser, G.C.H.E. de Croon (2015) Black-box lti modelling of flapping-wing micro aerial vehicle dynamics, in *AIAA conference 2015*
2. A. Briod, A. Klaptocz, J.-C. Zufferey, D. Floreano, The airburr: a flying robot that can exploit collisions, in *2012 ICME International Conference on Complex Medical Engineering (CME)* (2012), pp. 569–574, IEEE
3. A. Briod, K. Przemyslaw, J.-C. Zufferey, D. Floreano, A collision-resilient flying robot. J. Field Robot. **31**(4), 496–509 (2014)
4. P.-J. Bristeau, F. Callou, D. Vissiere, N. Petit et al., The navigation and control technology inside the ar. drone micro uav, in *18th IFAC World Congress*, vol. 18, pp. 1477–1484 (2011)
5. J.V. Caetano, C.C. de Visser, G.C.H.E. de Croon, B. Remes, C. de Wagter, J. Verboom, M. Mulder, Linear aerodynamic model identification of a flapping wing mav based on flight test data. Int. J. Micro Air Veh. **5**(4), 273–286 (2013)
6. J.V. Caetano, M. Percin, C.C. de Visser, B. van Oudheusden, G.C.H.E. de Croon, C. de Wagter, B. Remes, M. Mulder, Tethered vs. free flight force determination of the delfly ii flapping wing micro air vehicle, in *2014 International Conference on Unmanned Aircraft Systems (ICUAS)* (2014), pp. 942–948, IEEE

7. J.V. Caetano, M. Percin, B.W. van Oudheusden, B. Remes, C. de Wagter, G.C.H.E. de Croon, C.C. de Visser (2013) Unsteady forces acting on a flapping wing micro air vehicle: free-flight versus wind tunnel experimental methods (Submitted)

8. G.C.H.E. de Croon, K.M.E. de Clerq, R. Ruijsink, B. Remes, C. de Wagter, Design, aerodynamics, and vision-based control of the delfly. Int. J. Micro Air Veh. **1**(2), 71–97 (2009)

9. G.C.H.E. de Croon, M.A. Groen, C. de Wagter, B.D.W. Remes, R. Ruijsink, B.W. van Oudheusden, Design, aerodynamics, and autonomy of the delfly. Bioinspiration Biomimetics **7**(2), 025003 (2012)

10. J.W. Kruyt, E.M. Quicazán-Rubio, G.J.F. van Heijst, D.L. Altshuler, D. Lentink, Hummingbird wing efficacy depends on aspect ratio and compares with helicopter rotors. J. R. Soc. Interface **11**(99), 20140585 (2014)

11. F. Scarano, Tomographic PIV: principles and practice. Measur. Sci. Tech. **24**(1), 012001 (2013)

12. K.Y.W. Scheper, S. Tijmons, C.C. de Visser, G.C.H.E. de Croon, Behaviour trees for evolutionary robotics. Artif. Life **22**(1), (2015)

DelFly Versions

See Figs. A.1, A.2, A.3, A.4 and A.5.

© Springer Science+Bussiness Media Dordrecht 2016
G.C.H.E. de Croon et al., *The DelFly*, DOI 10.1007/978-94-017-9208-0

DelFly I

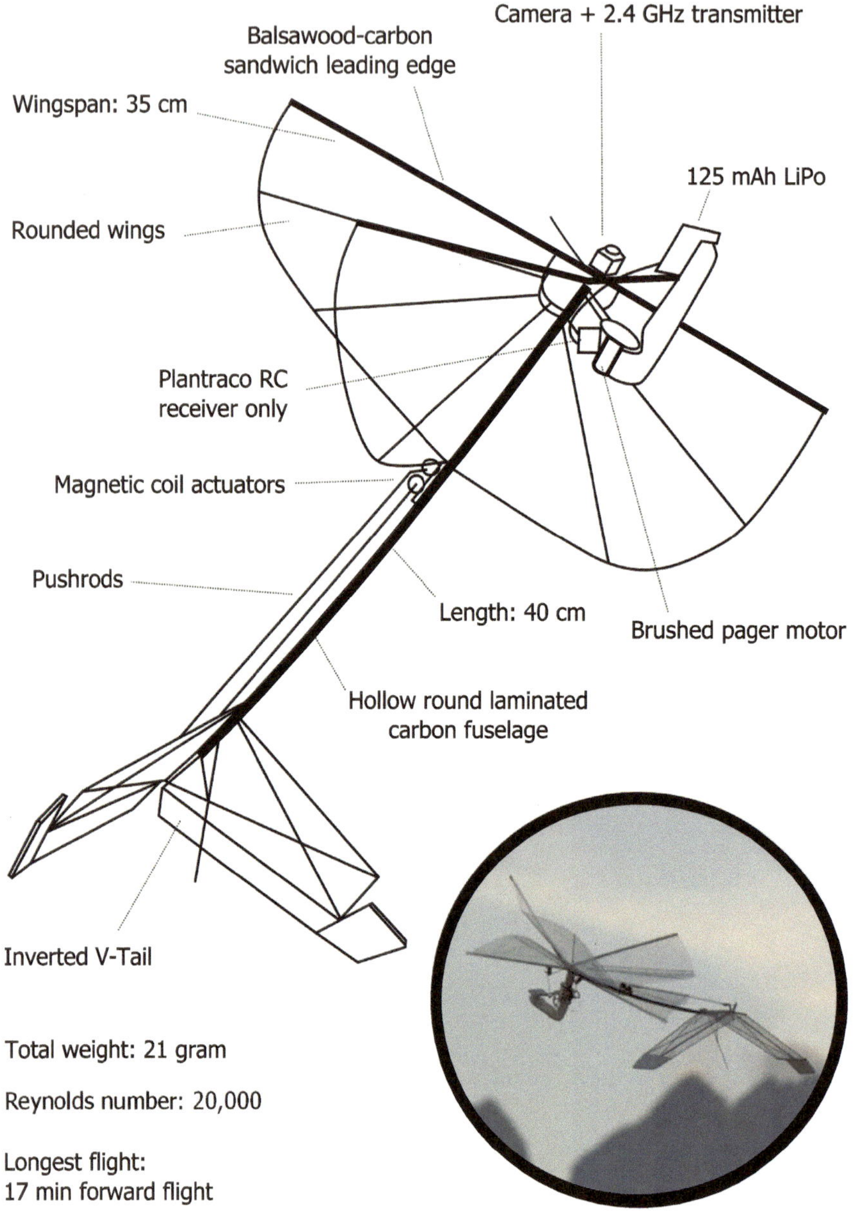

Balsawood-carbon
sandwich leading edge

Camera + 2.4 GHz transmitter

Wingspan: 35 cm

125 mAh LiPo

Rounded wings

Plantraco RC
receiver only

Magnetic coil actuators

Pushrods

Length: 40 cm

Brushed pager motor

Hollow round laminated
carbon fuselage

Inverted V-Tail

Total weight: 21 gram

Reynolds number: 20,000

Longest flight:
17 min forward flight

Fig. A.1 DelFly I: Average specifications of the DelFly I. The Reynolds number is determined on the basis of the mean wingtip velocity (Eq. 5.6). Only 5 were built and specifications like endurance also varied with aging. Reprinted with permission

DelFly II   *2007*

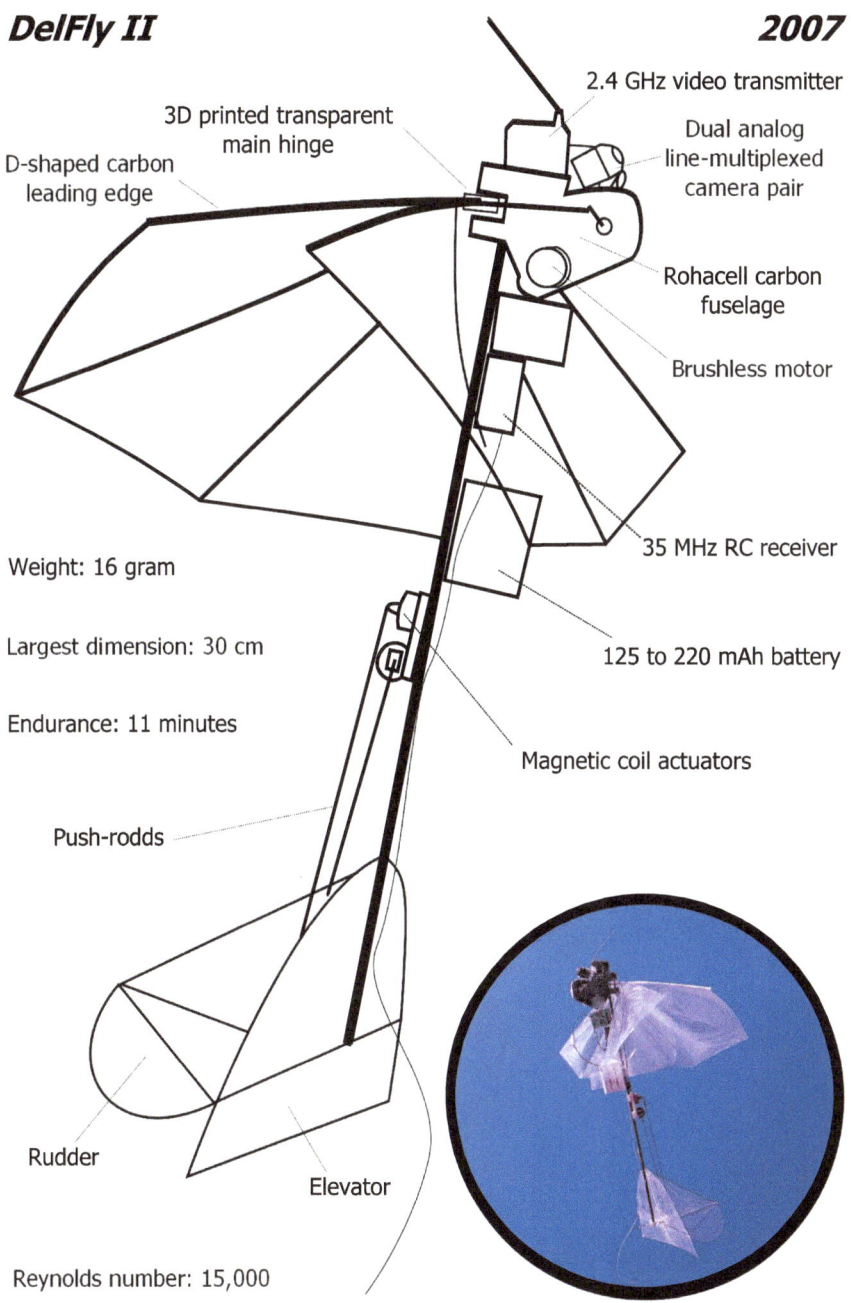

2.4 GHz video transmitter

3D printed transparent main hinge

Dual analog line-multiplexed camera pair

D-shaped carbon leading edge

Rohacell carbon fuselage

Brushless motor

Weight: 16 gram

35 MHz RC receiver

Largest dimension: 30 cm

125 to 220 mAh battery

Endurance: 11 minutes

Magnetic coil actuators

Push-rodds

Rudder

Elevator

Reynolds number: 15,000

Fig. A.2 Early DelFly II: Average specifications of the early DelFly II. Encircled photo by Jaap Oldenkamp. Reprinted with permission

DelFly II   *2009*

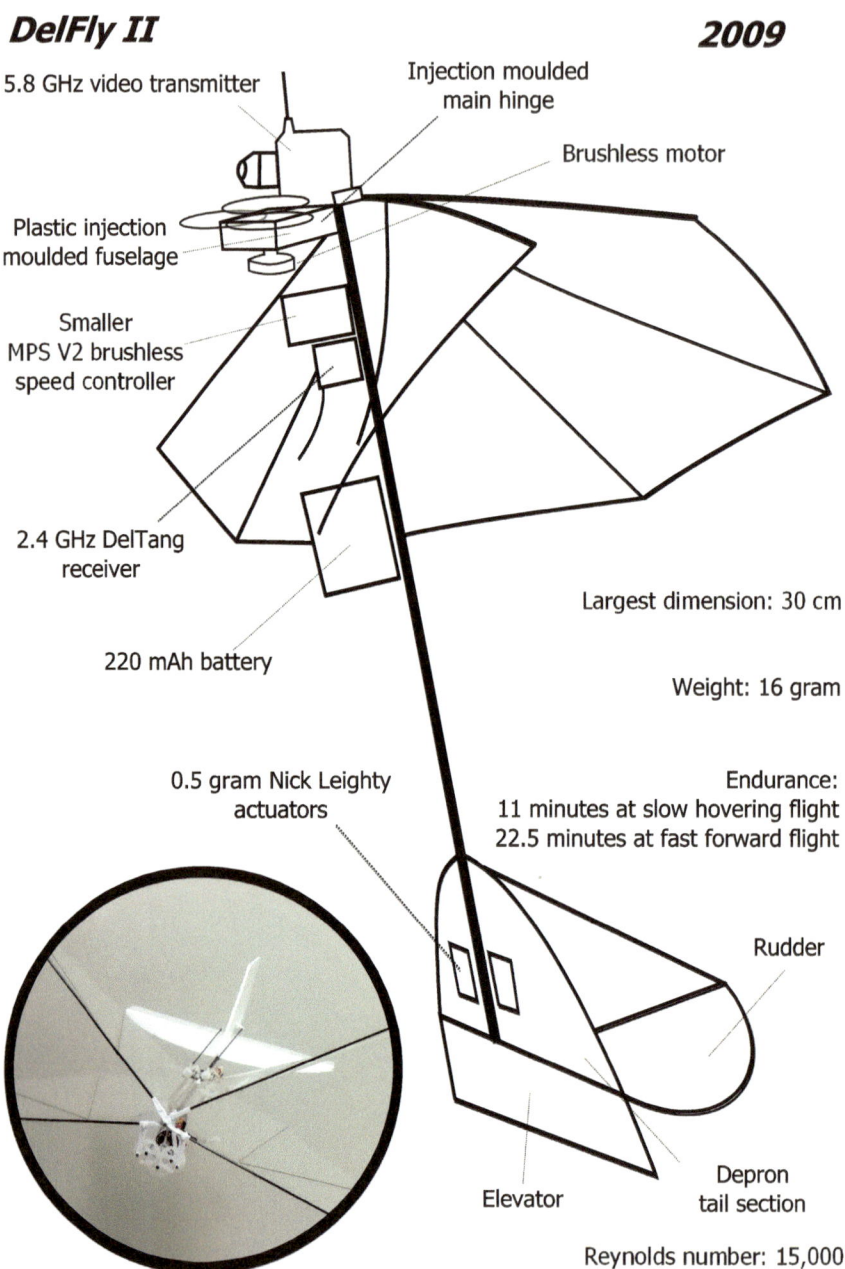

5.8 GHz video transmitter

Injection moulded
main hinge

Brushless motor

Plastic injection
moulded fuselage

Smaller
MPS V2 brushless
speed controller

2.4 GHz DelTang
receiver

Largest dimension: 30 cm

220 mAh battery

Weight: 16 gram

0.5 gram Nick Leighty
actuators

Endurance:
11 minutes at slow hovering flight
22.5 minutes at fast forward flight

Rudder

Depron
tail section

Elevator

Reynolds number: 15,000

Fig. A.3 Recent DelFly II: With injection moulded and extruded parts, the reproducibility of construction was significantly improved. But due to the large variety of research performed on DelFly II, specifications also vary. Typical values are given in this figure. Reprinted with permission

DelFly Micro

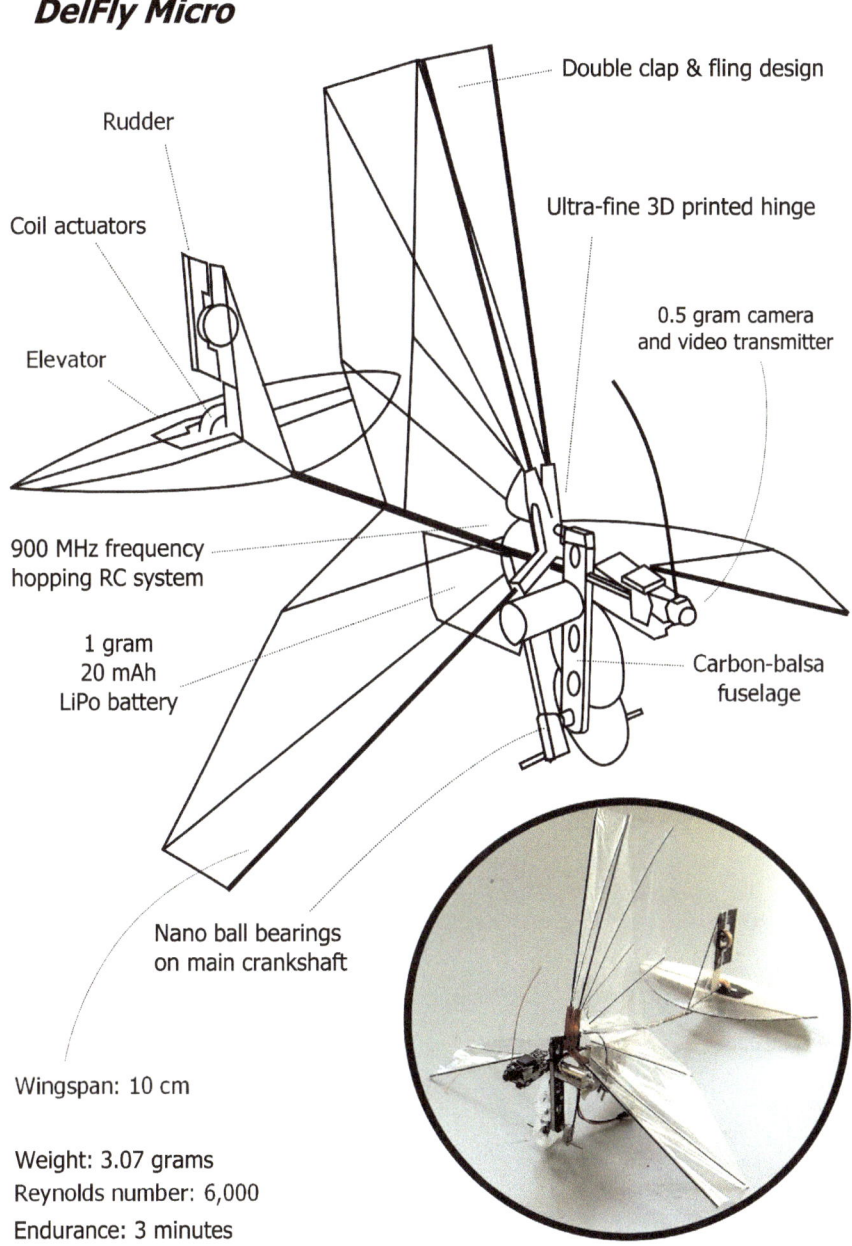

Double clap & fling design

Rudder

Ultra-fine 3D printed hinge

Coil actuators

0.5 gram camera
and video transmitter

Elevator

900 MHz frequency
hopping RC system

1 gram
20 mAh
LiPo battery

Carbon-balsa
fuselage

Nano ball bearings
on main crankshaft

Wingspan: 10 cm

Weight: 3.07 grams
Reynolds number: 6,000
Endurance: 3 minutes

Fig. A.4 DelFly Micro: After several unsuccessful prototypes, a single DelFly Micro was built with the given specifications. Reprinted with permission

DelFly Explorer

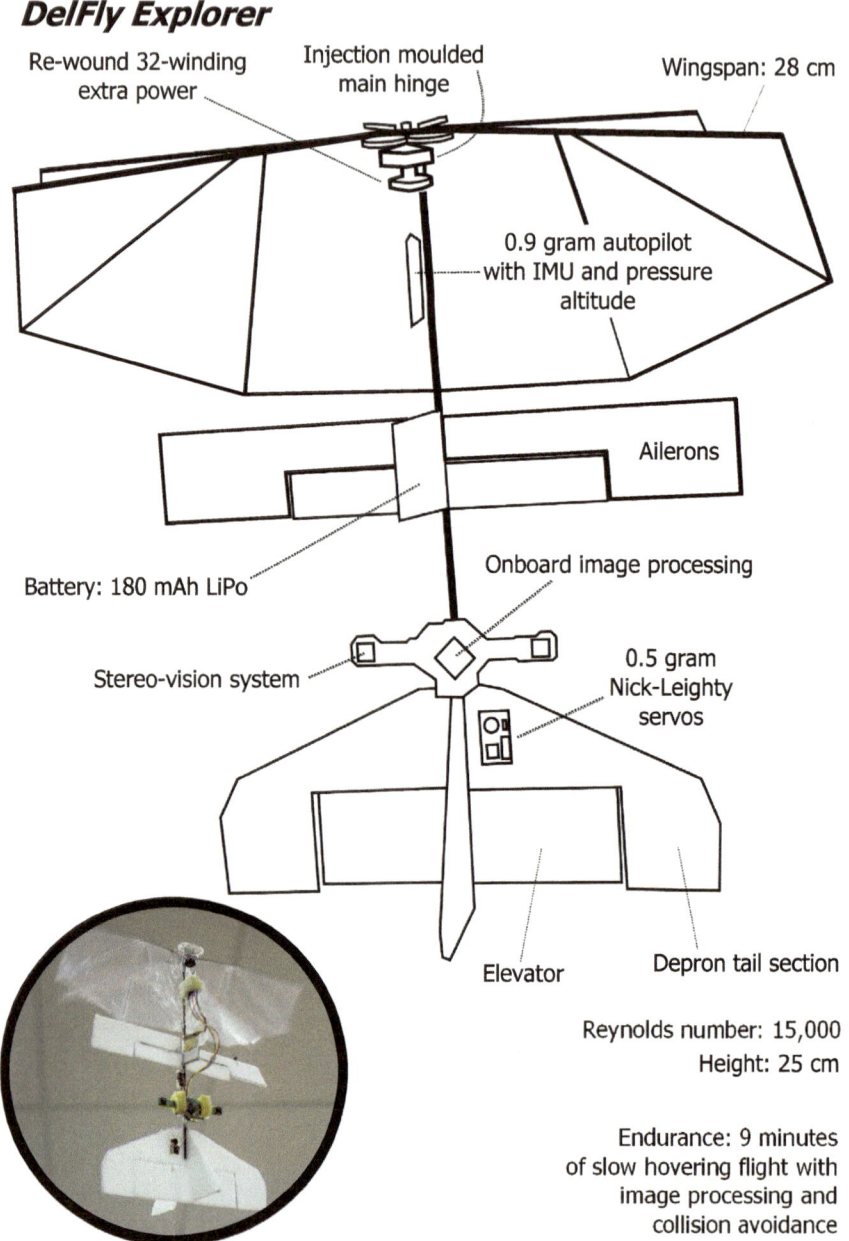

Re-wound 32-winding extra power

Injection moulded main hinge

Wingspan: 28 cm

0.9 gram autopilot with IMU and pressure altitude

Ailerons

Battery: 180 mAh LiPo

Onboard image processing

Stereo-vision system

0.5 gram Nick-Leighty servos

Elevator

Depron tail section

Reynolds number: 15,000
Height: 25 cm

Endurance: 9 minutes of slow hovering flight with image processing and collision avoidance

Fig. A.5 DelFly Explorer: Based on a DelFly II parts but packed with electronics and improved control and power, the DelFly Explorer can autonomously navigate in unknown indoor environments avoiding obstacles with onboard stereo vision. Reprinted with permission

Glossary

AI	Artificial Intelligence.
AUC	Area Under the Curve.
CDM	Color Distribution Method.
CFRP	Carbon Fiber Reinforced Polymer.
CMOS	Complementary MetalOxideSemiconductor.
CPLD	Complex Programmable Logic Device.
DARPA	Defense Advanced Research Projects Agency.
EA	Evolutionary Algorithm.
EMAV	European Micro Air Vehicle conference and competitions.
EMD	Elementary Motion Detector.
ER	Evolutionary Robotics.
ESC	Electronic Speed Controller.
EMF	Electromotive force.
FoE	Focus of Expansion.
FOV	Field Of View.
FPGA	Field Programmable Gate Array.
FSM	Finite State Machine.
FWMAV	Flapping Wing Micro Air Vehicle.
GPS	Global Positioning System.
HSV	Hue Saturation Value (image space).
IMAV	International Micro Air Vehicle conference and competitions.
IMU	Inertial Measurement Unit.
LED	Light-Emitting Diode.
LEV	Leading Edge Vortex.
MAV	Micro Air Vehicle.
MEMS	Microelectromechanical systems.
MSL	Mean Sea Level.
NAV	Nano Air Vehicle.
NTSC	Analog camera system named after National Television System Committee.

© Springer Science+Bussiness Media Dordrecht 2016
G.C.H.E. de Croon et al., *The DelFly*, DOI 10.1007/978-94-017-9208-0

PET	Polyethylene terephthalate.
PID	Proportional Integral Derivative.
PIV	Particle Image Velocimetry.
PWM	Pulse Width Modulation.
RAM	Random Access Memory.
ROC	Receiver Operator Characteristic.
RPM	Revolutions Per Minute.
SLAM	Simultaneous Localization And Mapping.
SMA	Shape Memory Alloy.
SNR	Signal to Noise Ratio.
TEV	Trailing Edge Vortex.
TMG	Texton Method with a Gray-scale dictionary.
TMC	Texton Method with a Color dictionary.
TTC	Time To Contact.
UAV	Unmanned Air Vehicle.

Index

© Springer Science+Bussiness Media Dordrecht 2016
G.C.H.E. de Croon et al., *The DelFly*, DOI 10.1007/978-94-017-9208-0